PLANT EXTRACTS IN NEURODEGENERATIVE DISEASES

PLANT EXTRACTS IN NEURODEGENERATIVE DISEASES

MAGISETTY OBULESU

Regional Agricultural Research Station, Acharya N.G. Ranga Agricultural University, Tirupati, India

ELSEVIER

ACADEMIC PRESS

An imprint of Elsevier

Academic Press is an imprint of Elsevier
125 London Wall, London EC2Y 5AS, United Kingdom
525 B Street, Suite 1650, San Diego, CA 92101, United States
50 Hampshire Street, 5th Floor, Cambridge, MA 02139, United States
The Boulevard, Langford Lane, Kidlington, Oxford OX5 1GB, United Kingdom

ISBN 978-0-323-95762-5

For information on all Academic Press publications
visit our website at https://www.elsevier.com/books-and-journals

Publisher: Nikki P Levy
Acquisitions Editor: Natalie Farra
Editorial Project Manager: Czarina Mae S Osuyos
Production Project Manager: Selvaraj Raviraj
Cover Designer: Christian J Bilbow

Typeset by STRAIVE, India

Dedication

This work is dedicated to my beloved parents, Magisetty Jagan Mohan and Magisetty Saraswathi, my wife, and all my family members (sisters, daughter, and son). I had praiseworthy mental support during this project and throughout my life.

Contents

About the author

Magisetty Obulesu is a Scientist in UR Advanced Therapeutics, Hyderabad. He was a Research Associate in the Department of Materials Science, Graduate School of Pure and Applied Sciences at the University of Tsukuba. He worked as a Research Associate at the Regional Agricultural Research Station, Acharya N.G. Ranga Agricultural University, Tirupati, India. He has 22 years of research and teaching experience. He is also on the editorial boards of several nanotechnology journals including *SciFed Nanotech Research Letters, SciFed Drug Delivery Research, Current Updates in Nanotechnology*, and *Journal of Nanotechnology and Materials Science*. His research areas include food science, pathology of neurodegenerative diseases such as Alzheimer's disease, designing polymer-based biomaterials such as hydrogels, and development of metal chelators to overcome metal-induced toxicity. He has researched Alzheimer's disease and developed an aluminum-induced neurotoxicity rabbit model. Mr. Obulesu's present research focuses on development of redox-active injectable hydrogels of polyion complex. He has written three books and edited two others; he is also the first and corresponding author for most of his articles. He was awarded the Tsukuba Scholarship and Nikki-Saneyoshi Scholarship in Japan and Senior Research Fellowship in India from the Indian Council of Medical Research (ICMR). He was also awarded a certificate from Stanford University for their Scientific Writing course.

Acknowledgments

I sincerely acknowledge the endeavors of my ever-memorable teacher, Mr. Nageswara Rao, Director of the Apex Institute of English, Guntur, Andhra Pradesh, India. His worth-commending didactic skills made my English language skills reach culminating points. I also acknowledge all my teachers and mentors who significantly nurtured my research and writing skills.

CHAPTER 1

Effect of plant extracts against Alzheimer's disease

Introduction

Pathology

Alzheimer's disease (AD) is an age-linked disease that more commonly affects the elderly. Usually, AD diagnosis before age 65 is significantly low and found in a small percentage of people (2%–5% of all cases) possessing genetic mutations in corresponding genes (Bekris et al., 2010). Although AD is found in families, with genetic approach on one hand but surpassing the role of genetics on other hand, mounting evidence has shown that the genetic risk factors account for only one-third of brain modifications that occur with age (Small et al., 2000; Cole et al., 2019). The other two-thirds of nongenetic factors probably are based on lifestyle and the environment.

The APOE gene on chromosome 19, such as the APOE4 gene, encodes a protein that shows a crucial role in cholesterol metabolism. Nevertheless, in contrast to amyloid precursor protein (APP) and presenilins, this gene merely signifies a risk factor for late-onset AD in 60% of cases (Bird, 2008; Liu et al., 2013; Zhou et al., 2019).

Amyloid beta (Aβ) and tau start aggregating in the brain many years before the diagnosis of clinical symptoms. Age-associated plaques are noticed in brain areas including the hippocampus, amygdala, and neocortex (Kodali et al., 2015; Subramaniam, 2019). The Aβ peptide is a small peptide obtained from the proteolytic breakdown of APP by β-secretase and γ-secretase through the secretory amyloidogenic pathway occurring in several neuronal sections, such as axons, nerve terminals, and dendrites (Yamin et al., 2008; Poddar et al., 2019). These pathological events occur several years before the onset of the characteristic plaques, and the accumulation of Aβ takes place many years before the progression of clinical dementia and can be ante-mortem as shown by PET-amyloid imaging of the brain of AD subjects (Rodrigue et al., 2009; Rowe and Villemagne, 2011; Serrano-Pozo et al., 2011).

Plant Extracts in Neurodegenerative Diseases
https://doi.org/10.1016/B978-0-323-95762-5.00001-1

In addition, studies on postmortem brain tissue have demonstrated that a decrease in the activity of a few enzymes of the tricarboxylic acid cycle is noticed in AD patients (Sorbi et al., 1983; Butterworth and Besnard, 1990; Mastrogiacoma et al., 1996). Bubber et al. (2005) studied a total evaluation of the activity of the complete enzymes of the Krebs cycle in AD brain tissue and corroborated the reduction in the activity of pyruvate dehydrogenase and α-ketoglutarate dehydrogenase complexes, together with a profound decrease in the isocitrate dehydrogenase activity. Conversely, the succinate dehydrogenase and malate dehydrogenase enzymes exhibited enhanced activity. The role of oxidative phosphorylation (OXPHOS) complexes is ambiguous, although the functional irregularities of the Krebs cycle and the respiratory chain certainly stimulate a modification of the energy metabolism and result in an enhanced generation of reactive oxygen species (ROS), thus making the mitochondrial membrane permeable and provoking programmed cell death (Bubber et al., 2005; Atlante et al., 2017). Since the etiopathogenesis of AD is complex, the utilization of incredible natural compounds to treat neurodegeneration in AD is very essential (Mancuso et al., 2012; Rahman et al., 2017; Nasrullah et al., 2017; Uddin et al., 2018a,b, 2019a,b,c).

Plant extracts

Diet

Diet enriched with vegetables and fruits offer several health benefits, according to voluminous epidemiological studies (Bergamini, 2010; Atlante et al., 2020).

Diet also has an ability to induce disease. Healthy eating averts AD, and interestingly, a diet that keeps Alzheimer's at bay is nearly similar to what makes the heart healthy, lowers cholesterol, prevents cancer, and balances glucose levels (Brown, 2015; Crimmins, 2015). Indeed, green leafy vegetables and fruits present significant benefits to health, but several other foods do substantially improve the human brain. Therefore, it can be concluded that exact dietary suggestions for AD patients are currently gaining ground (Barnard et al., 2014; Cremonini et al., 2019; Amini et al., 2020).

In general, the existence of high oxidizable content like lipid milieu of the myelin membrane of neurons render the brain vulnerable to oxidative injury. Consequently, antioxidant foods play a vital neuroprotective role in neural function (Teleanu et al., 2019; Cenini et al., 2019; Singh et al., 2019). In line with this, multifarious berries exert robust antioxidant activity due to the occurrence of tannins, anthocyanins, and phenols that substantially enhance the plasticity of the hippocampus, therefore, initiating learning and memory function. Alpha lipoic acid, abundantly available in vegetables like spinach and broccoli, also contributes toward regulating the energy homeostasis of mitochondria and ameliorating cognitive ability. Additionally, green and black tea, both enriched with antioxidants, possess epigallocatechin gallate, which was observed to indirectly mitigate the build-up of amyloid plaques, a pathological characteristic of AD. Eggs are rich in several nutrients such as vitamin B6, vitamin B12, and choline and folic acid, and may contribute to brain health (Subash et al., 2014; Cascella et al., 2017; Colizzi, 2018; Dos Santos et al., 2019; Simunkova et al., 2019; Moretti and Peinkhofer, 2019). Turmeric, a CUR-rich spice that imparts the yellow color to curry, curtails memory impairment induced by AD by impeding the synthesis of amyloid plaques. Red wine enriched with resveratrol (RSV) substantially combats neurodegeneration (Mazzanti and Di Giacomo, 2016; Caruana et al., 2016; Reale et al., 2020).

Apoptosis also plays a pivotal role in AD contributing to the death of huge neuronal populations. In addition, one of the causative factors that could initiate an unusual activation of the self-removal program for apoptosis of complete neuronal regions would alarm neurotrophins. These molecules, which possess the nerve growth factor (NGF) as its progenitor, primarily curtails the neuronal death program; thus, a neuronal population devoid of the basal source of particular neurotrophins suffers enormous apoptosis (Calissano et al., 1998; Ryu et al., 2016; Fricker et al., 2018).

Polyphenols

The polyphenols, including carotenoids and a few bioactive compounds, are categorized under the plant kingdom as substances termed as "functional," since they induce substantial health benefits (Singh, 2018). They are profoundly focused as useful tools to overcome AD and also to impede disease progression (Singh, 2018). A plethora of compounds playing a role as precursors of other compounds are required in neuronal metabolism and brain health regulation. For example, mono-unsaturated fatty acids (MUFA) and polyunsaturated fatty acids (PUFA, ω-3 and ω-6), enriched in fish and vegetable oils and also the B vitamins, folic acid, vitamins B6, and B12, are substances required for healthy neuronal functioning and for their advantages on cognitive and behavioral efficiency (Rathod et al., 2016; Kennedy, 2016).

Flavonoids

Several lines of evidence have shown that flavonoids protect from AD by intervening with the generation and accumulation of Aβ peptides and/or reducing the build-up of tau. Flavonoids are capable of eliminating Aβ peptides and attenuate tau phosphorylation by the mTOR/autophagy signaling pathway. In addition, because of their cholinesterase suppressive activity, flavonoids can be substantial anti-AD compounds (Uddin et al., 2020a,b). In numerous diseases such as neurodegenerative disease (ND), diabetes, and cancer, natural compounds serve as a substantial source for a variety of molecular characteristics, biochemical specificity, and massive chemical multiplicity, which renders these natural products suitable for the alteration of several signaling cascades (Rasul et al., 2013; Uddin et al., 2017, 2018a,b, 2020b). Flavonoids are usually available in multifarious vegetables, fruits, and plants (Liu et al., 2014; Uddin and Upaganlawar, 2019). These natural compounds are recognized to have a broad range of pharmacological actions (Uddin et al., 2019b; Uddin and Upaganlawar, 2019) and also act as robust metal chelators, free radical scavengers, and antioxidant agents (Uddin et al., 2019; Zhang et al., 2015; Elbaz et al., 2016; Tysnes and Storstein, 2017). Flavonoids also pacify microglial activation to regulate inflammatory processes in the central nervous system (CNS) (Spencer et al., 2012), harbor robust antiamyloidogenic, antidepressant effects (Nabavi et al., 2015), and ameliorate memory and learning efficacy (Kim et al., 2009). Moreover, these natural compounds demonstrate antiinflammatory (Li et al., 2010; Ashafaq et al., 2012; Ejaz Ahmed et al., 2013), neuroprotective (Prakash and Sudhandiran, 2015; Gomes et al., 2015), antiaging (Lin et al., 2015), and anticholinesterase (Khan et al., 2018) activities.

Polyphenols, primarily flavonoids, are profoundly found as flavanones in citrus fruits (Mecocci et al., 2014) and isoflavones in soy (Francis et al., 2006; Wang et al., 2014), and a few polyphenols like flavan-3-ols (also known as flavanols) exhibit remarkable health benefits (Francis et al., 2006). They are catechin, epicatechin, epigallocatechin, and epigallocatechin gallate available in several vegetable products, such as cocoa, chocolate, black and green tea, and grapes. Mitochondrial impairment and oxidative stress, which contribute to neural membrane damage and memory dysfunction (Uttara et al., 2009; Jacob et al., 2013; Wang et al., 2014; Saharan and Mandal, 2014; Tönniesa and Trushinaa, 2017; Gomes et al., 2018; Wang et al., 2020), are biochemical characteristics observed in AD.

Mitochondria, impaired due to enhanced oxidative stress, enhance ROS generation and Aβ peptides, which in turn initiates oxidative stress both in vitro and in vivo, resulting in neuronal apoptosis and eventually contributing to AD progression (Gomes et al., 2018). In the human body, autophagy eliminates injured cells and facilitates the regeneration of healthy and newer cells (Uddin et al., 2019c; Tanjir Islam et al., 2017). In addition, autophagy also regulates the generation and removal of Aβ (Nilsson and Saido, 2014). With a view toward decreasing synaptic imperfection and neuronal death, elimination of Aβ from the brain is another primary target for anti-AD drugs (Lukiw, 2012).

Genistein

Genistein, an isoflavone primarily available in soy products, substantially combats AD (Mukund et al., 2017). Existence of a plethora of phenolic moieties in its structure makes genistein induce robust antioxidant properties (Sadhukhan et al., 2018). Hence, genistein has antioxidant and neuroprotective efficacy against in vitro AD models (Park et al., 2016). It mitigates oxidative stress by suppressing the production of ROS; it also rescues mitochondria by enhancing the reduced-to-oxidized glutathione ratio and decreasing 8-oxo-20-deoxyguanosine, an indicator of mitochondrial DNA damage (Devi et al., 2017). A study carried out on primary cultured neurons investigated the neuroprotective effect on cells treated with the Aβ fragment 31–35 (Aβ31–35).

Voluminous data has shown that the short fragment Aβ31–35, derived by proteolytic cleavage of Aβ precursor protein (APP), could induce apoptosis in primary cultured neurons (Yu et al., 2009) and PC12 cells (Zhang et al., 2006), serving as an "active center" of the full length Aβ molecule with robust neuron toxicity (Yu et al., 2009; Clementi et al., 2005). Ding et al. (2011) found that genistein treatment in neurons, 2h before inclusion of the Aβ 31–35 fragment, has been found to protect neurons by enhancing cell viability and decreasing both Ca^{2+} and ROS production. In addition, the ratio of GSH/GSSG in mitochondria and mitochondrial membrane potential have also been enhanced after genistein treatment. Pierzynowska et al. (2019) proved that an increased dose (i.e., 150 mg/kg/day) of genistein induced activation of autophagy in a streptozotocin-induced rat model of the sporadic AD. Furthermore, at this high dose, it has also been observed that genistein provoked the complete breakdown of Aβ and hyperphosphorylated tau by initiating autophagy.

Icariside

Icariside, an extract of *Herba epimedii*, is one of the earliest herbal medicines. Aggregates of Aβ contribute to the progression in AD, and the aggregation of these aggregates initiate toxicity in the brain. The primary controlling alpha-secretase in neurons, A disintegrin and metalloproteinase 10 (ADAM10), play a role in breaking down APP. This cleavage produces a neuroprotective APP-obtained petite fragment (C83) and inhibits the generation of Aβ peptides (Endres and Deller, 2017). The mechanistic role of icariside in AD therapy includes enhanced ADAM10 expression and reduced expression of both APP and BACE1, leading to decreased generation of Aβ (Yan et al., 2017; Wuli et al., 2020).

Onjisaponin

Onjisaponin B (OB) reduces Aβ by cleavage of APP without suppressing the beta-amyloid-cleaving enzyme (BACE1) and γ-secretase activities. In vivo, oral introduction of OB curtails Aβ pathology and behavioral deficits in APP/PS1 mice (Singh et al., 2015; Li et al., 2016a,b). OB can decrease IL-1β, IL-6, TNF-α, and malondialdehyde in the serum and hippocampus of lipopolysaccharide-initiated cognitive-impaired rats. OB induces antioxidant, antiinflammatory, and antiapoptosis protective efficacy in lipopolysaccharide-treated PC12 cells (Li et al., 2018, Fig. 1).

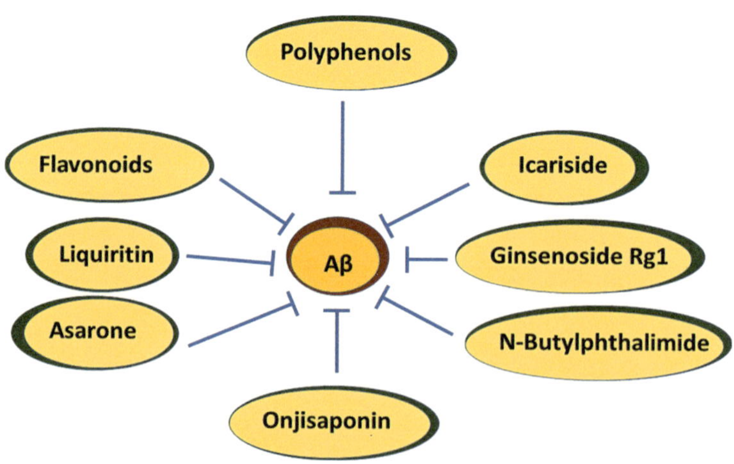

Fig. 1 This image shows the protective efficacy of multifarious plant bioactive compounds against AD.

Asarones

Asarones, the essential ingredients of *Acori Tatarinowii Rhizoma*, can ameliorate the proliferation and self-restoration of neural progenitor cells. Studies have proved that asarones activated extracellular signal-regulated kinase (ERK), a fragment of a critical kinase cascade for neurogenesis (Mao et al., 2015). Asarones have reduced Aβ accumulation in the cortex and hippocampus of APP/PS1 mice brains (Yang et al., 2016, Fig. 1).

Liquiritin

Liquiritin has been isolated from *Glycyrrhiza uralensis*. As shown before, oxidative stress is a vital event in the pathogenesis of AD. Aβ peptide is a histopathological hallmark of AD. Aβ can deposit in association with metal ions to catalyze ROS generation (Cheignon et al., 2018). Enhanced oxidative stress instigate pathology. Malondialdehyde (MDA) and 8-hydroxy-20 deoxyguanosine (8-OHdG) are the resultant molecules of lipid peroxidation and are extensively found biomarkers of oxidative stress (Draper and Hadley, 1990). Liquiritin lowers MDA and 8-OHdG in the hippocampus of rats with AD and, by relevant mechanisms, substantially ameliorates spatial learning and memory (Jia et al., 2016).

Tanshinone IIA and cryptotanshinone

Tanshinone IIA (TIIA) and cryptotanshinone (CT) are isolated from *Salviae miltiorrhizae* as chief lipophilic compounds. In vitro, TIIA and CT markedly mitigated intracellular NO and Ca^{2+} levels in H9c2 cells (Jin and Li, 2013). The gliosis-linked and neuroinflammatory compounds in the hippocampal tissues showed a significant decrease in the expression of glial fibrillary acidic protein (GFAP), S100β protein, COX-2, inducible nitric oxide synthase (iNOS), and nuclear factor kappa-light-chain-enhancer of activated B cells (NF-kBp65). TIIA and CT exert antiinflammatory and neuroprotective efficacy in a nongenetic mouse model of AD (Maione et al., 2018).

Ginsenoside Rg1

Panax ginseng is the source of Ginsenoside Rg1. It ameliorated hippocampal long-term potentiation and memory in an AD mouse model. It primarily enhances the activity of antioxidant enzymes such as glutathione peroxidase and superoxide dismutase, curtails activation, and eventually enhances the proliferation of hippocampal neurons (Li et al., 2016a,b; Zhu et al., 2014).

n-Butylidenephthalide

n-Butylidenephthalide (*n*-BP) is obtained from *Angelica sinensis*. It is used in the treatment of NDs such as Parkinson's disease (Fu et al., 2014), amyotrophic lateral sclerosis (Hsueh et al., 2016), and AD (Chang et al., 2015). *n*-BP was observed to decrease Aβ, total tau protein and its hyperphosphorylated form, and cellular toxicity in iPSCs-derived neurons stimulated by Down syndrome (Chang et al., 2015).

Green nanotechnology

Plant-based production of NPs is a simple method where a metal salt is mixed with a plant extract and the reaction is carried out in a short time at room temperature. Using this strategy, silver (Ag) and gold (Au) NPs were produced that are robust, corresponding to other metallic NPs (Gour and Jain, 2019). Synthesis of silver nanoparticles (SNPs) with multifarious medicinal plants for pharmaceutical and biological applications has long been in use (Mani et al., 2012). Green-synthesized NPs induce therapeutic applications, e.g., antimicrobial (Rajenran et al., 2015), antioxidant (Mohan et al., 2014), cytotoxic (Shawkey et al., 2013), and antiinflammatory properties (Mani et al., 2015). Nanotechnology has been of significant importance in the diagnosis and treatment of AD currently (Nazem and Mansoori, 2011). Numerous NPs including titanium dioxide, silica dioxide, and silver and zinc oxide showed therapeutic efficacy against neurological diseases (Nazıroğlu et al., 2017), where oxidative NPs can reduce the activities of ROS-scavenging enzymes like glutathione peroxidase (GSH-Px), superoxide dismutase (SOD), and catalase in the brain of rats and mice (Nazıroğlu et al., 2017; Youssif et al., 2019).

Conclusions and future perspectives

Since the discovery of AD by Alois Alzheimer, numerous plant extracts and their bioactive ingredients have been extensively studied for their protective efficacy against it. Plant extracts showed exemplary protection against various pathological events and hallmarks, such as Aβ and tau, through various mechanistic roles. However, low bioavailability and short systemic circulation impede their success. To overcome these challenges, nanotechnology has been used extensively. More specifically, use of green nanotechnology that comprises only natural compounds play an essential role in AD treatment. Despite the success of numerous plant-based molecules and nanotechnology-based drug delivery systems, intensive further studies are warranted to corroborate these findings.

References

Amini, Y., Saif, N., Greer, C., Hristov, H., Isaacson, R., 2020. The role of nutrition in individualized Alzheimer's risk reduction. Curr. Nutr. Rep. 9, 55–63.

Ashafaq, M., Raza, S.S., Khan, M.M., Ahmad, A., Javed, H., Ahmad, M.E., et al., 2012. Catechin hydrate ameliorates redox imbalance and limits inflammatory response in focal cerebral ischemia. Neurochem. Res. 37, 1747–1760.

Atlante, A., de Bari, L., Bobba, A., Amadoro, G., 2017. A disease with a sweet tooth: exploring the Warburg effect in Alzheimer's disease. Biogerontology 18, 301–319.

Atlante, A., Amadoro, G., Bobba, A., Latina, V., 2020. Functional foods: an approach to modulate molecular mechanisms of Alzheimer's disease. Cell 9, 2347.

Barnard, N.D., Bush, A.I., Ceccarelli, A., Cooper, J., de Jager, C.A., Erickson, K.I., et al., 2014. Dietary and lifestyle guidelines for the prevention of Alzheimer's disease. Neurobiol. Aging 35, S74–S78.

Bekris, L.M., Yu, C.E., Bird, T.D., Tsuang, D.W., 2010. Genetics of Alzheimer disease. J. Geriatr. Psychiatry Neurol. 23, 213–227.

Bergamini, E., 2010. Nutraceuticals: a valuable aid to be used cautiously. G. Gerontol. 58, 255–258.

Bird, T.D., 2008. Genetic aspects of Alzheimer disease. Genet. Med. 10, 231–239.

Brown, G.C., 2015. Living too long: the current focus of medical research on increasing the quantity, rather than the quality, of life is damaging our health and harming the economy. EMBO Rep. 16, 137–141.

Bubber, P., Haroutunian, V., Fisch, G., Blass, J.P., Gibson, G.E., 2005. Mitochondrial abnormalities in Alzheimer brain: mechanistic implications. Ann. Neurol. 57, 695–703.

Butterworth, R.F., Besnard, A.M., 1990. Thiamine-dependent enzyme changes in temporal cortex of patients with Alzheimer's disease. Metab. Brain Dis. 5, 179–184.

Calissano, P., Ciotti, M.T., Galli, C., Mercanti, D., Dus, L., Canu, N., et al., 1998. The role of IGF-I in cerebellar granule cell survival and terminal differentiation. In: Müller, E.E. (Ed.), IGFs in the Nervous System. 1998. Springer, Berlin, Germany, pp. 60–71.

Caruana, M., Cauchi, R., Vassallo, N., 2016. Putative role of red wine polyphenols against brain pathology in Alzheimer's and Parkinson's disease. Front. Nutr. 3, 31.

Cascella, M., Bimonte, S., Muzio, M.R., Schiavone, V., Cuomo, A., 2017. The efficacy of Epigallocatechin-3-gallate (green tea) in the treatment of Alzheimer's disease: an overview of pre-clinical studies and translational perspectives in clinical practice. Infect. Agent Cancer 12, 36.

Cenini, G., Lloret, A., Cascella, R., 2019. Oxidative stress in neurodegenerative diseases: from a mitochondrial point of view. Oxid. Med. Cell. Longev. 2019, 18.

Chang, C.Y., Chen, S.M., Lu, H.E., Lai, S.M., Lai, P.S., Shen, P.W., Chen, P.Y., Shen, C.I., Harn, H.J., Lin, S.Z., et al., 2015. N-butylidenephthalide attenuates Alzheimer's disease-like cytopathy in Down syndrome induced pluripotent stem cell-derived neurons. Sci. Rep. 5, 8744.

Cheignon, C., Tomas, M., Bonnefont-Rousselot, D., Faller, P., Hureau, C., Collin, F., 2018. Oxidative stress and the amyloid beta peptide in Alzheimer's disease. Redox Biol. 14, 450–464.

Clementi, M.E., Marini, S., Coletta, M., Orsini, F., Giardina, B., Misiti, F., 2005. Abeta(31–35) and Abeta(25–35) fragments of amyloid beta-protein induce cellular death through apoptotic signals: role of the redox state of methionine-35. FEBS Lett. 579, 2913–2918.
Cole, J.H., Marioni, R.E., Harris, S.E., Deary, I.J., 2019. Brain age and other bodily 'ages': implications for neuropsychiatry. Mol. Psychiatry 24, 266–281.
Colizzi, C., 2018. The protective effects of polyphenols on Alzheimer's disease: a systematic review. Alzheimers Dement. 5, 184–196.
Cremonini, A.L., Caffa, I., Cea, M., Nencioni, A., Odetti, P., Monacelli, F., 2019. Nutrients in the prevention of Alzheimer's disease. Oxid. Med. Cell. Longev. 2019, 20.
Crimmins, E.M., 2015. Lifespan and healthspan: past, present, and promise. Gerontologist 55, 901–911.
Devi, K.P., Shanmuganathan, B., Manayi, A., Nabavi, S.F., Nabavi, S.M., 2017. Molecular and therapeutic targets of genistein in Alzheimer's disease. Mol. Neurobiol. 54, 7028–7041.
Ding, B., Yuan, L., Yu, H., Li, L., Ma, W., Bi, X., et al., 2011. Genistein and folic acid prevent oxidative injury induced by β-amyloid peptide. Basic Clin. Pharmacol. Toxicol. 108, 333–340.
Dos Santos, S.M., Romeiro, C.F.R., Rodrigues, C.A., Cerqueira, A.R.L., Monteiro, M.C., 2019. Mitochondrial dysfunction and alpha-lipoic acid: beneficial or harmful in Alzheimer's disease? Oxid. Med. Cell. Longev. 2019, 14.
Draper, H.H., Hadley, M., 1990. Malondialdehyde determination as index of lipid peroxidation. Methods Enzymol. 186, 421–431.
Ejaz Ahmed, M., Khan, M.M., Javed, H., Vaibhav, K., Khan, A., Tabassum, R., et al., 2013. Amelioration of cognitive impairment and neurodegeneration by catechin hydrate in rat model of streptozotocin-induced experimental dementia of Alzheimer's type. Neurochem. Int. 62, 492–501.
Elbaz, A., Carcaillon, L., Kab, S., Moisan, F., 2016. Epidemiology of Parkinson's disease. Rev. Neurol. 172, 14–26.
Endres, K., Deller, T., 2017. Regulation of alpha-secretase ADAM10 in vitro and in vivo: genetic, epigenetic and protein-based mechanisms. Front. Mol. Neurosci. 10, 56.
Francis, S.T., Head, K., Morris, P.G., Macdonald, I.A., 2006. The effect of flavanol-rich cocoa on the fMRI response to a cognitive task in healthy young people. J. Cardiovasc. Pharmacol. 47 (Suppl. 2), S215–S220.
Fricker, M., Tolkovsky, A.M., Borutaite, V., Coleman, M., Brown, G.C., 2018. Neuronal cell death. Physiol. Rev. 98, 813–880.
Fu, R.H., Harn, H.J., Liu, S.P., Chen, C.S., Chang, W.L., Chen, Y.M., et al., 2014. n-Butylidenephthalide protects against dopaminergic neuron degeneration and alpha-synuclein accumulation in Caenorhabditis elegans models of Parkinson's disease. PLoS One 9, e85305.
Gomes, A., Pimpão, R.C., Fortalezas, S., Figueira, I., Miguel, C., Aguiar, C., et al., 2015. Chemical characterization and bioactivity of phytochemicals from Iberian endemic Santolina semidentata and strategies for ex situ propagation. Ind. Crop Prod. 74, 505–513.
Gomes, B.A.Q., Silva, J.P.B., Romeiro, C.F.R., Dos Santos, S.M., Rodrigues, C.A., Gonçalves, P.R., et al., 2018. Neuroprotective mechanisms of resveratrol in Alzheimer's disease: role of SIRT1. Oxid. Med. Cell. Longev. 2018, 15.
Gour, A., Jain, N., 2019. Advances in green synthesis of nanoparticles. Artif. Cells Nanomed. Biotechnol. 47 (1), 844–851.
Hsueh, K.W., Chiou, T.W., Chiang, S.F., Yamashita, T., Abe, K., Borlongan, C.V., et al., 2016. Autophagic down-regulation in motor neurons remarkably prolongs the survival of ALS mice. Neuropharmacology 108, 152–160.
Jacob, K.D., Hooten, N.N., Trzeciak, A.R., Evans, M.K., 2013. Markers of oxidant stress that are clinically relevant in aging and age-related disease. Mech. Ageing Dev. 134, 139–157.

Jia, S.L., Wu, X.L., Li, X.X., Dai, X.L., Gao, Z.L., Lu, Z., et al., 2016. Neuroprotective effects of liquiritin on cognitive deficits induced by soluble amyloid-beta1-42 oligomers injected into the hippocampus. J. Asian Nat. Prod. Res. 18, 1186–1199.

Jin, H.J., Li, C.G., 2013. Tanshinone IIA and cryptotanshinone prevent mitochondrial dysfunction in hypoxia-induced H9c2 cells: association to mitochondrial ROS, intracellular nitric oxide and calcium levels. Evid. Based Complement. Alternat. Med. 2013, 610694.

Kennedy, D.O., 2016. B vitamins and the brain: mechanisms, dose and efficacy—a review. Nutrients 8, 68.

Khan, H., Marya, Amin, S., Kamal, M.A., Patel, S., 2018. Flavonoids as acetylcholinesterase inhibitors: current therapeutic standing and future prospects. Biomed. Pharmacother. 101, 860–870.

Kim, D.H., Kim, S., Jeon, S.J., Son, K.H., Lee, S., Yoon, B.H., et al., 2009. Tanshinone I enhances learning and memory, and ameliorates memory impairment in mice via the extracellular signal-regulated kinase signalling pathway. Br. J. Pharmacol. 158, 1131–1142.

Kodali, M., Parihar, V.K., Hattiangady, B., Mishra, V., Shuai, B., Shetty, A.K., 2015. Resveratrol prevents age-related memory and mood dysfunction with increased hippocampal neurogenesis and microvasculature and reduced glial activation. Sci. Rep. 28, 8075.

Li, Q., Zhao, H., Zhao, M., Zhang, Z., Li, Y., 2010. Chronic green tea catechins administration prevents oxidative stress-related brain aging in C57BL/6J mice. Brain Res. 1353, 28–35.

Li, X., Cui, J., Yu, Y., Li, W., Hou, Y., Wang, X., et al., 2016a. Traditional Chinese nootropic medicine radix polygalae and its active constituent onjisaponin B reduce beta-amyloid production and improve cognitive impairments. PLoS One 11, e0151147.

Li, F., Wu, X., Li, J., Niu, Q., 2016b. Ginsenoside Rg1 ameliorates hippocampal long-term potentiation and memory in an Alzheimer's disease model. Mol. Med. Rep. 13, 4904–4910.

Li, X., Sun, Y., Wei, Y., Zhou, L., Liu, L., Yin, P., et al., 2018. Onjisaponin B (OB) is neuroprotective during cognitive loss through immune-mediated and SIRT1 pathways. Curr. Neurovasc. Res. 15, 94–102.

Lin, L., Ni, B., Lin, H., Zhang, M., Li, X., Yin, X., et al., 2015. Traditional usages, botany, phytochemistry, pharmacology and toxicology of Polygonum multiflorum Thunb.: a review. J. Ethnopharmacol. 159, 158–183.

Liu, C.C., Kanekiyo, T., Xu, H., Bu, G., 2013. Apolipoprotein E and Alzheimer disease: risk, mechanisms, and therapy. Nat. Rev. Neurol. 9, 106–118.

Liu, C.M., Ma, J.Q., Liu, S.S., Zheng, G.H., Feng, Z.J., Sun, J.M., 2014. Proanthocyanidins improves lead-induced cognitive impairments by blocking endoplasmic reticulum stress and nuclear factor-κB-mediated inflammatory pathways in rats. Food Chem. Toxicol. 72, 295–302.

Lukiw, W.J., 2012. Amyloid beta (Aβ) peptide modulators and other current treatment strategies for Alzheimer's disease (AD). Expert Opin. Emerg. Drugs. https://doi.org/10.1517/14728214.2012.672559.

Maione, F., Piccolo, M., De Vita, S., Chini, M.G., Cristiano, C., De Caro, C., et al., 2018. Down regulation of pro-inflammatory pathways by tanshinone IIA and cryptotanshinone in a non-genetic mouse model of Alzheimer's disease. Pharmacol. Res. 129, 482–490.

Mancuso, C., Siciliano, R., Barone, E., Preziosi, P., 2012. Natural substances and Alzheimer's disease: from preclinical studies to evidence-based medicine. Biochim. Biophys. Acta Mol. basis Dis. 1822, 616–624.

Mani, A., Lakshmi, S., Gopal, V., 2012. Bio-mimetic synthesis of silver nanoparticles and evaluation of its free radical scavenging activity. Int. J. Biol. Pharm. Res. 3 (4), 631–633.

Mani, A., Seethalakshmi, S., Gopal, V., 2015. Evaluation of *in vitro* anti-inflammatory activity of silver nanoparticles synthesised using *piper nigrum* extract. J. Nanomedicine Nanotechnol. 6, 2.

Mao, J., Huang, S., Liu, S., Feng, X.L., Yu, M., Liu, J., et al., 2015. A herbal medicine for Alzheimer's disease and its active constituents promote neural progenitor proliferation. Aging Cell 14, 784–796.

Mastrogiacoma, F., Bettendor, L., Grisar, T., Kish, S.J., 1996. Brain thiamine, its phosphate esters, and its metabolizing enzymes in Alzheimer's disease. Ann. Neurol. 39, 585–591.

Mazzanti, G., Di Giacomo, S., 2016. Curcumin and resveratrol in the management of cognitive disorders: what is the clinical evidence? Molecules 21, 1243.

Mecocci, P., Tinarelli, C., Schulz, R.J., Polidori, M.C., 2014. Nutraceuticals in cognitive impairment and Alzheimer's disease. Front. Pharmacol. 5, 147.

Mohan, S., Sasikala, K., Anand, T., Vengaiah, P., Krishnaraj, S., 2014. Green synthesis, antimicrobial and antioxidant effects of silver nanoparticles using *Canthium coromandelicum* leaves extract. Res. J. Microbiol. 9 (3), 142–150.

Moretti, R., Peinkhofer, C., 2019. B vitamins and fatty acids: what do they share with small vessel disease-related dementia? Int. J. Mol. Sci. 20, 5797.

Mukund, V., Mukund, D., Sharma, V., Mannarapu, M., Alam, A., 2017. Genistein: its role in metabolic diseases and cancer. Crit. Rev. Oncol. Hematol. 119, 13–22.

Nabavi, S.F., Braidy, N., Habtemariam, S., Orhan, I.E., Daglia, M., Manayi, A., et al., 2015. Neuroprotective effects of chrysin: from chemistry to medicine. Neurochem. Int. 90, 224–231.

Nasrullah, M., Haque, A., Alzahrani, S.A., Ashraf Uddin, M., Yasmin, Z., Sahab Uddin, M., et al., 2017. Acetylcholinesterase and butyrylcholinesterase enzyme inhibitory effect of *Alangium salviifolium* (L. f.) Wang pericarp extracts with their phytochemical and antioxidant values. J. Pharm. Res. Int. 19, 1–11.

Nazem, A., Mansoori, G., 2011. Nanotechnology for Alzheimer's disease detection and treatment. Insciences J. 4, 169–193.

Nazıroğlu, M., Muhamad, S., Pecze, L., 2017. Nanoparticles as potential clinical therapeutic agents in Alzheimer's disease: focus on selenium nanoparticles. Expert. Rev. Clin. Pharmacol. 10, 751–2441.

Nilsson, P., Saido, T.C., 2014. Dual roles for autophagy: degradation and secretion of Alzheimer's disease Aβ peptide. Bioessays 36, 570–578.

Park, Y.J., Ko, J.W., Jeon, S., Kwon, Y.H., 2016. Protective effect of genistein against neuronal degeneration in $ApoE^{-/-}$ mice fed a high-fat diet. Nutrients 8, 692.

Pierzynowska, K., Podlacha, M., Gaffke, L., Majkutewicz, I., Mantej, J., Węgrzyn, A., et al., 2019. Autophagy-dependent mechanism of genistein-mediated elimination of behavioral and biochemical defects in the rat model of sporadic Alzheimer's disease. Neuropharmacology 148, 332–346.

Poddar, J., Pradhan, M., Ganguly, G., Chakrabarti, S., 2019. Biochemical deficits and cognitive decline in brain aging: intervention by dietary supplements. J. Chem. Neuroanat. 95, 70–80.

Prakash, D., Sudhandiran, G., 2015. Dietary flavonoid fisetin regulates aluminium chloride-induced neuronal apoptosis in cortex and hippocampus of mice brain. J. Nutr. Biochem. 26, 1527–1539.

Rahman, A., Haque, A., Uddin, M.S., Mian, M.M., Sufian, M.A., Rahman, M.M., et al., 2017. In vitro screening for antioxidant and anticholinesterase effects of Uvaria littoralis Blume.: a nootropic phytotherapeutic remedy. J. Intellect. Disabil. Diagnosis Treat. 5, 50–60.

Rajenran, R., Ganesan, N., Balu, S.K.A., Thandavamorthy, P., Thiruvengadam, D., 2015. Green synthesis, characterization, antimicrobial and cytotoxic effects of silver nanoparticles using Origanum heracleoticum L. leaf extract. Int. J. Pharm. Pharm. Sci. 7 (4), 288–293.

Rasul, A., Di, J., Millimouno, F., Malhi, M., Tsuji, I., Ali, M., et al., 2013. Reactive oxygen species mediate isoalantolactone-induced apoptosis in human prostate cancer cells. Molecules 18, 9382–9396.

Rathod, R., Kale, A., Joshi, S., 2016. Novel insights into the effect of vitamin B12 and omega-3 fatty acids on brain function. J. Biomed. Sci. 23, 17.

Reale, M., Costantini, E., Jagarlapoodi, S., Khan, H., Belwal, T., Cichelli, A., 2020. Relationship of wine consumption with Alzheimer's disease. Nutrients 12, 206.

Rodrigue, K.M., Kennedy, K.M., Park, D.C., 2009. Beta-amyloid deposition and the aging brain. Neuropsychol. Rev. 19, 436–450.

Rowe, C.C., Villemagne, V.L., 2011. Brain amyloid imaging. J. Nucl. Med. 52, 1733–1740.

Ryu, J.R., Hong, C.J., Kim, J.Y., Kim, E.K., Sun, W., Yu, S.W., 2016. Control of adult neurogenesis by programmed cell death in the mammalian brain. Mol. Brain 9, 43.

Sadhukhan, P., Saha, S., Dutta, S., Mahalanobish, S., Sil, P.C., 2018. Nutraceuticals: an emerging therapeutic approach against the pathogenesis of Alzheimer's disease. Pharmacol. Res. 129, 100–114.

Saharan, S., Mandal, P.K., 2014. The emerging role of glutathione in Alzheimer's disease. J. Alzheimers Dis. 40, 519–529.

Serrano-Pozo, A., Frosch, M.P., Masliah, E., Hyman, B.T., 2011. Neuropathological alterations in Alzheimer disease. Cold Spring Harb. Perspect. Med. 1, a006189.

Shawkey, A., Rabeh, M., Abdulall, A., Abdellatif, A., 2013. Green nanotechnology: anticancer activity of silver nanoparticles using *Citrullus colocynthis* aqueous extracts. Adv. Life Sci. Technol. 3, 60–70.

Simunkova, M., Alwasel, S.H., Alhazza, I.M., Jomova, K., Kollar, V., Rusko, M., et al., 2019. Management of oxidative stress and other pathologies in Alzheimer's disease. Arch. Toxicol. 93, 2491–2513.

Singh, R., 2018. Current Alzheimer's management with berries fruits therapy. J. Public Health Nutr. 1, 17–24.

Singh, J.C., Kakalij, R.M., Kshirsagar, R.P., Kumar, B.H., Komakula, S.S., Diwan, P.V., 2015. Cognitive effects of vanillic acid against streptozotocin-induced neurodegeneration in mice. Pharm. Biol. 53, 630–636.

Singh, A., Kukreti, R., Saso, L., Kukreti, S., 2019. Oxidative stress: a key modulator in neurodegenerative diseases. Molecules 24, 1583.

Small, G.W., Ercoli, L.M., Silverman, D.H., Huang, S.C., Komo, S., Bookheimer, S.Y., et al., 2000. Cerebral metabolic and cognitive decline in persons at genetic risk for Alzheimer's disease. Proc. Natl. Acad. Sci. U. S. A. 97, 6037–6042.

Sorbi, S., Bird, E.D., Blass, J.P., 1983. Decreased pyruvate dehydrogenase complex activity in Huntington and Alzheimer brain. Ann. Neurol. 13, 72–78.

Spencer, J.P.E., Vafeiadou, K., Williams, R.J., Vauzour, D., 2012. Neuroinflammation: modulation by flavonoids and mechanisms of action. Mol. Aspects Med. 33, 83–97.

Subash, S., Essa, M.M., Al-Adawi, S., Memon, M.A., Manivasagam, T., Akbar, M., 2014. Neuroprotective effects of berry fruits on neurodegenerative diseases. Neural Regen. Res. 9, 1557–1566.

Subramaniam, S., 2019. Selective neuronal death in neurodegenerative diseases: the ongoing mystery. Yale J. Biol. Med. 92, 695–705.

Tanjir Islam, M., Sahab Uddin, M., Nahar Lucky, K., Manjurul Islam, M., Md Saiful Islam, S., Tanvir Kabir, M., et al., 2017. Autophagic dysfunction in type 2 diabetes mellitus: pathophysiology and therapeutic implications. J. Diabetes Metab. 8, 1–8.

Teleanu, R.I., Chircov, C., Grumezescu, A.M., Volceanov, A., Teleanu, D.M., 2019. Antioxidant therapies for neuroprotection-a review. J. Clin. Med. 8, 1659.

Tönniesa, E., Trushinaa, E., 2017. Oxidative stress, synaptic dysfunction, and Alzheimer's disease. J. Alzheimers Dis. 57, 1105–1121.

Tysnes, O.B., Storstein, A., 2017. Epidemiology of Parkinson's disease. J. Neural Transm. 124, 901–905.

Uddin, M.S., Upaganlawar, A., 2019. Oxidative Stress and Antioxidant Defense Biomedical Value in Health and Diseases. Nova Science Publishers, Hauppauge, NY, USA. ISBN 9781536156874.

Uddin, M.S., Uddin, G.M.S., Begum, M.M., Begum, Y., Herrera-Calderon, O., Islam, M.M., et al., 2017. Inspection of phytochemical content and in vitro antioxidant profile of Gnaphalium luteoalbum L.: an unexplored phytomedicine. J. Pharm. Nutr. Sci. 7, 136–146.

Uddin, M.S., Hossain, M.S., Kabir, M.T., Rahman, I., Tewari, D., Jamiruddin, M.R., et al., 2018a. Phytochemical screening and antioxidant profile of Syngonium podophyllum schott stems: a fecund phytopharmakon. J. Pharm. Nutr. Sci. 8, 120–128.

Uddin, M.S., Hossain, M.S., Mamun, A.A., Tewari, D., Asaduzzaman, M., Islam, M.S., et al., 2018b. Phytochemical analysis and antioxidant profile of methanolic extract of seed, pulp and peel of Baccaurea ramiflora Lour. Asian Pac. J. Trop. Med. 11, 443–450.

Uddin, M.S., Al Mamun, A., Kabir, M.T., Jakaria, M., Mathew, B., Barreto, G.E., et al., 2019a. Nootropic and anti-Alzheimer's actions of medicinal plants: molecular insight into therapeutic potential to alleviate Alzheimer's neuropathology. Mol. Neurobiol. 56, 4925–4944.

Uddin, M.S., Kabir, M.T., Tewari, D., Mathew, B., Aleya, L., 2019b. Emerging signal regulating potential of small molecule biflavonoids to combat neuropathological insults of Alzheimer's disease. Sci. Total Environ. 700, 1–11.

Uddin, M.S., Mamun, A.A., Labu, Z.K., Hidalgo-Lanussa, O., Barreto, G.E., Ashraf, G.M., 2019c. Autophagic dysfunction in Alzheimer's disease: cellular and molecular mechanistic approaches to halt Alzheimer's pathogenesis. J. Cell. Physiol. 234, 8094–8112.

Uddin, M., Kabir, M., Niaz, K., Jeandet, P., Clement, C., Mathew, B., et al., 2020a. Molecular insight into the therapeutic promise of flavonoids against Alzheimer's disease. Molecules 25, 1267.

Uddin, M.S., Mamun, A.A., Jakaria, M., Thangapandiyan, S., Ahmad, J., Rahman, M.A., et al., 2020b. Emerging promise of sulforaphane-mediated Nrf2 signaling cascade against neurological disorders. Sci. Total Environ. 707, 1–12.

Uttara, B., Singh, A.V., Zamboni, P., Mahajan, R.T., 2009. Oxidative stress and neurodegenerative diseases: a review of upstream and downstream antioxidant therapeutic options. Curr. Neuropharmacol. 7, 65–74.

Wang, X., Wang, W., Li, L., Perry, G., Lee, H., Zhua, X., 2014. Oxidative stress and mitochondrial dysfunction in Alzheimer's disease. Biochim. Biophys. Acta 1842, 1240–1247.

Wang, W., Zhao, F., Ma, X., Perry, G., Zhu, X., 2020. Mitochondria dysfunction in the pathogenesis of Alzheimer's disease: recent advances. Mol. Neurodegener. 15, 1–22.

Wuli, W., Tsai, S., Chiou, T., Harn, H., 2020. Human-induced pluripotent stem cells and herbal small-molecule drugs for treatment of Alzheimer's disease. Int. J. Mol. Sci. 21, 1327.

Yamin, G., Ono, K., Inayathullah, M., Teplow, D.B., 2008. Amyloid β-protein assembly as a therapeutic target of Alzheimer's disease. Curr. Pharm. Des. 14, 3231–3246.

Yan, L., Deng, Y., Gao, J., Liu, Y., Li, F., Shi, J., et al., 2017. Icariside II effectively reduces spatial learning and memory impairments in Alzheimer's disease model mice targeting beta-amyloid production. Front. Pharmacol. 8, 106.

Yang, C., Li, X., Mo, Y., Liu, S., Zhao, L., Ma, X., et al., 2016. beta-Asarone mitigates amyloidosis and downregulates RAGE in a transgenic mouse model of Alzheimer's disease. Cell. Mol. Neurobiol. 2016 (36), 121–130.

Youssif, K.A., Haggag, E.G., Elshamy, A.M., Rabeh, M.A., Gabr, N.M., Seleem, A., et al., 2019. Anti-Alzheimer potential, metabolomic profiling and molecular docking of green synthesized silver nanoparticles of *Lampranthus coccineus* and *Malephora lutea* aqueous extracts. PLoS One 14, e0223781.

Yu, H.L., Li, L., Zhang, X.H., Xiang, L., Zhang, J., Feng, J.F., et al., 2009. Neuroprotective effects of genistein and folic acid on apoptosis of rat cultured cortical neurons induced by β-amyloid 31–35. Br. J. Nutr. 102, 655–662.

Zhang, J.M., Wu, M.N., Qi, J.S., Qiao, J.T., 2006. Amyloid beta-protein fragment 31–35 suppresses long-term potentiation in hippocampal CA1 region of rats in vivo. Synapse 60, 307–313.

Zhang, Q., Zhao, J.J., Xu, J., Feng, F., Qu, W., 2015. Medicinal uses, phytochemistry and pharmacology of the genus Uncaria. J. Ethnopharmacol. 173, 48–80.

Zhou, X., Chen, Y., Mok, K.Y., Kwok, T.C.Y., Mok, V.C.T., Guo, Q., Alzheimer's Disease Neuroimaging Initiative, et al., 2019. Non-coding variability at the APOE locus contributes to the Alzheimer's risk. Nat. Commun. 10, 3310.

Zhu, J., Mu, X., Zeng, J., Xu, C., Liu, J., Zhang, M., et al., 2014. Ginsenoside Rg1 prevents cognitive impairment and hippocampus senescence in a rat model of D-galactose-induced aging. PLoS One 9, e101291.

CHAPTER 2

Potential plant extracts in the treatment of Parkinson's disease

Introduction

Pathology

Parkinson's disease (PD) is a progressive, devastating neurodegenerative disease (ND), onset of which includes continuous loss of dopaminergic neurons in the substantia nigra pars compacta (SNc) (Herrera et al., 2017). It is generally an age-associated movement disorder usually observed sporadically (Collier et al., 2017). The pathogenesis of PD has been elusive, although a plethora of studies exhibit a range of genetic and environmental factors in the onset and development of PD (Cannon and Greenamyre, 2013). An array of pathological events of PD are mitochondrial dysfunction, excitotoxicity, endoplasmic reticulum stress, oxidative/nitrosative stress, and inflammation, as well as ubiquitin-proteasome system impairment (Moore et al., 2005; Lashuel et al., 2013; Ur Rasheed et al., 2016; Gelders et al., 2018). Overall, such pathological events stimulate the aggregation of anomalous or misfolded α-synuclein (α-syn) protein (Moore et al., 2005; Lashuel et al., 2013; Ur Rasheed et al., 2016; Gelders et al., 2018). A wealth of genetic, biochemical, cellular, pathological, and molecular studies showed that PD pathology is linked to environments where α-syn is vulnerable to polymerization, aggregation, and fibril formation and propagation (Moore et al., 2005; Hansen and Li, 2012; Lashuel et al., 2013; Gelders et al., 2018; Ghiglieri et al., 2018). The α-syn oligomers induce mitochondrial impairment and provoke endoplasmic stress, oxidative stress, and neuroinflammation, and inhibit proteasomal activity and autophagy (Ghiglieri et al., 2018).

α-Synuclein

α-Syn, a 140-residue presynaptic protein, is involved in the trafficking and fusion of synaptic vesicles, and regulates dopamine release at presynaptic terminals (Burre et al., 2010; Bendor et al., 2013). In general, the human brain has α-syn in the concentration of 1 μM in the brain and 70 pM in

Plant Extracts in Neurodegenerative Diseases
https://doi.org/10.1016/B978-0-323-95762-5.00003-5

cerebrospinal fluid (Borghi et al., 2000). It is present in natively unfolded monomer form and achieves an α-helical secondary structure on associating with lipid vesicles. When destabilized, misfolding is followed by accumulation of α-syn in neurons (Ruipérez et al., 2010; Bartels et al., 2011; Broersen et al., 2018). Monomeric α-syn is a fundamentally disordered protein observed in various conformational states. It shows a remarkable role in numerous essential biochemical processes (Tompa, 2005) and in an increasing number of diseases characterized by misfolding, particularly PD (Uversky and Dunker, 2010). α-Syn is a presynaptic protein that plays a pivotal role in the secretion of neurotransmitters through synaptic vesicles of the brain. α-Syn aggregates such as Lewy bodies are structures of both sporadic and familial forms of PD. These aggregates experience many important stages of fibrillation, oligomerization, and aggregation (Javed et al., 2019).

Numerous molecules such as antibodies (Bergström et al., 2016), polyamines (Büttner et al., 2014), heat shock proteins (Cox et al., 2016), chaperones (Friesen et al., 2017), and pharmaceuticals (Lauterbach et al., 2010) were observed to interact with various forms of α-syn (i.e., monomers, soluble oligomers, protofibrils, or fibrils) oligomerization, fibrillation, and promote scavenging. Consequently, targeting α-syn aggregation, oligomerization, fibrillation, and propagation to inhibit α-syn toxicity has been regarded as a vital therapeutic tool for reducing or hampering disease development (Kalia et al., 2015; Török et al., 2016; Brundin et al., 2017).

Plant extracts

A plethora of studies emphasized the neuroprotective efficacy of plant extracts and phytochemicals in PD by antioxidant and antiinflammatory activities (Sarrafchi et al., 2016; da Costa et al., 2017; Mazo et al., 2017; Morgan and Grundmann, 2017; Wang et al., 2017; Amro et al., 2018; Zhang et al., 2017). Although antioxidants (both natural and synthetic) showed significant protective efficacy in preclinical studies, coenzyme Q10 (Beal et al., 2014), creatine (Attia et al., 2017), and vitamin E (Ahlskog, 1994) did not show adequate neuroprotective efficacy in patients. Numerous natural products (primarily plant extracts and phytochemicals) were developed to target α-syn (Masuda et al., 2006; Meng et al., 2009, 2010a; Caruana et al., 2011; Marchiani et al., 2013).

A plethora of plant extracts demonstrated (in vitro) protective efficacy in experimental models of PD by targeting α-syn. Nevertheless, the bioactive ingredients inducing protective efficacy are hitherto unknown.

Bacopa monnieri curbs neurodegeneration in A53T α-syn-instigated PD in *Caenorhabditis elegans* (Jadiya et al., 2011). However, the chemical components, together called Bacosides, have not been studied in experimental PD models or efficacy on α-syn. *Centella asiatica* (L.) Urb., also called Asiatic pennywort, averts α-syn accumulation in vitro (Berrocal et al., 2014), although the vital component asiatic acid could not avert α-syn aggregation. The protective efficacy of asiaticoside and madecassic acid have not been examined against α-syn. Cinnamon extract precipitate has been found to attenuate α-syn aggregation and stabilization of oligomers in vitro and in vivo in A53T α-syn-initiated PD in flies (Shaltiel-Karyo et al., 2012). *Eukalyptus citriodora* ameliorated climbing efficacy and inhibited oxidative stress in transgenic drosophila expressing human α-syn (Siddique et al., 2013).

Saffron and its components like crocin-1, crocin-2, crocetin, safranal, and the crocetin structural analogs hexadecanedioic acid, norbixin, and *trans*-muconic acid successfully targeted α-syn fibrillation and aggregation (Inoue et al., 2018). Conversely, a few crocetin analogs could not target α-syn accumulation and separation. *Sorbus alnifolia*, familiar as Korean mountain ash, ameliorated viability of rat pheochromocytoma (PC12) cells in turn enhancing longevity and food sensing, and decreasing dopaminergic neurodegeneration in *Caenorhabditis elegans* model of PD (Cheon et al., 2017). Nevertheless, the extract did not change α-syn aggregation in the NL5901 strain (Cheon et al., 2017, Fig. 1).

Pharmacological chaperones combat Parkinson's disease

The advent of pharmacological chaperoning has been considered a substantial therapeutic avenue in several diseases linked to single gene mutations (Srinivasan et al., 2014). Chaperones are small molecules that attach to proteins, stabilize them, and protect from proteolytic breakdown or thermal

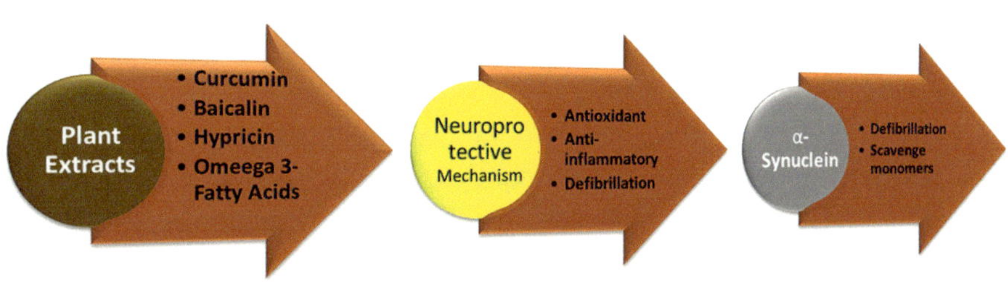

Fig. 1 The image shows protective efficacy of multifarious plant bioactive compounds against PD.

denaturation. In addition, they facilitate or avert a few protein-protein assemblies in line with the molecular chaperones (Ringe and Petsko, 2009). Manifold molecular chaperones like Hsp70, Hsp40, and torsin A impede the misfolding of proteins or initiate the breakdown and removal of misfolded proteins, thus opening a promising therapeutic avenue in PD (Dimant et al., 2012).

Despite the potential therapeutic role of molecular chaperoning in α-syn-mediated neurodegeneration, the structural heterogeneousness and deficit of determined structural constituents for α-syn remains a biological riddle of significance in the development of small molecules to combat α-syn (Lester et al., 2009). Plant-derived phytochaperones play a key role in targeting protein misfolding in neurotherapeutics (Bernd, 2008). In line with this, *Ginkgo biloba* is focused to discover bioactive molecules that can be used in drug discovery to overcome protein-misfolding diseases inducing neurodegeneration (Kastenholz and Garfin, 2009).

Phytochemicals

The phytochemicals are secondary metabolites without a nutritive role but with a substantial role as a source of drugs (Beutler, 2009; Henrich and Beutler, 2013). Among the various phytochemicals such as alkaloids, saponins, carotenoids, lignans, glycosides, etc., the alkaloids with nitrogen exert neuroprotection against NDs (Hussain et al., 2018). Several of the alkaloids like acetylcorynoline, 3a-acetoxyeudesma-1,4 (15),11 (13)-trien-12, 6a-olide, corynoxine B, DL-3-*n*-butylphthalide, isorhynchophylline, and squamosamide inhibit neurotoxicity in experimental models by suppressing α-syn aggregation or fibril formation.

Baicalein

Baicalein, a flavone extracted from roots of *Scutellaria baicalensis Georgi*, a well-known plant in traditional Chinese medicine, has been found to avert α-syn oligomerization and fibrillation (Sashourpour et al., 2017). It networks with α-syn via a tyrosine residue. After oxidation, it produces quinone metabolites that combine covalently with a lysine side chain in α-syn. It curbs fibril formation and also breaks down preformed fibrils at low micromolar concentrations (Zhu et al., 2004). Meng et al. (2009) showed that its noncovalent attachment with α-syn and covalent modification by the oxidized form limits the conformational modifications in the unfolded protein, thus stabilizing to α-syn monomers and oligomers. Baicalein separated the α-syn fibrillation states (monomeric, oligomeric intermediate, and fibrillar forms) with the help of a JC-1 probe and corroborated the fibrillation inhibitory activity of baicalein on α-syn (Lee et al., 2009). In PC12 cells, it encounters cytotoxicity, mitochondrial depolarization, and attenuates proteasome suppression stimulated by E46K, an α-syn point mutation similar to familial PD (Jiang et al., 2010).

Curcumin

Curcumin derived from turmeric, chemically called diferuloylmethane, plays a pivotal role in drug discovery and development (Molino et al., 2016). It has a significant dietary role and numerous health effects, and is an extensively investigated phytochemical in experimental and clinical studies (Molino et al., 2016). It exerts neuroprotection against NDs such as PD and shows antioxidant, antiinflammatory, and antiapoptotic properties (Ji and Shen, 2014, Obulesu, 2021). Ono and Yamada (2006) observed that curcumin has antifibrillogenic activity by suppressing α-syn fibril formation and destabilizing preformed fibrils (Ono and Yamada, 2006). It has been observed to attenuate oligomerization of mutant α-syn into high molecular weight aggregates (Pandey et al., 2008) and stimulate the separation of α-syn fibrils (Shoval et al., 2008).

Curcumin along with β-cyclodextrin demonstrated a cooperative suppression of α-syn accumulation and degenerated the preformed aggregates into monomers at lower concentrations (Gautam et al., 2014, 2017). Gautam et al. (2017) also showed that an appropriate placement of the phenolic groups, benzene rings, and flexibility potentiates the efficacy of curcumin. The phenolic groups improve curcumin's interactions with α-syn monomers and oligomers.

Nanotechnology

Curcumin has low stability, solubility, and oral bioavailability. With a view to enhance its stability, solubility, and oral bioavailability, numerous nanoformulations or structural analogs have been studied (Gadad et al., 2012; Kundu et al., 2016; Taebnia et al., 2016). Curc-gluc, an altered curcumin preparation, attenuated α-syn oligomerization and fibrillation (Gadad et al., 2012). In PC12 cells treated with H_2O_2, MPP+, and $MnCl_2$, dehydrozingerone, zingerone, and their biphenyl analogs exhibited cytoprotective efficacy (Marchiani et al., 2013). The biphenyl analogs of dehydrozingerone and *O*-methyl-dehydrozingerone avoid α-syn accumulation; the biphenyl zingerone analog has been found to be a robust inhibitor and has promising antioxidant activity. This activity has been induced due to the existence of the hydroxylated biphenyl scaffold in the pharmacophore (Marchiani et al., 2013).

Taebnia et al. (2016) synthesized amine-conjugated mesoporous silica nanoparticles of curcumin to improve its bioavailability and studied its protective efficacy against cytotoxicity and α-syn fibrillation (Taebnia et al., 2016). This nanoformulation demonstrated interaction with α-syn species and averted fibrillation with insignificant effect on cytotoxicity (Taebnia et al., 2016). A nanoformulation comprising curcumin and piperine with glyceryl monooleate nanoparticles coated with different surfactants was designed for targeted delivery to increase its bioavailability in the brain (Kundu et al., 2016).

The nanoformulation substantially inhibited oxidative stress, apoptosis, and averted α-syn oligomerization and fibrillation and stimulated autophagy. Nine curcumin analogs have been developed by replacement of groups on the aromatic ring that modified the hydrophobicity, initiated stability, and promoted binding with the fibrils and oligomers (Jha et al., 2016). A few of the analogs enhanced stability and seemed to unite with oligomers and preformed fibrils.

Scutellaria pinnatifida (*S. pinnatifida*)

For many years, herbal medicines have been in use for a wide range of disorders, and interest in them has been increasing tremendously. Accordingly, multifarious species of the *Scutellaria* genus have been focused for pharmaceutical studies because of their repository of secondary metabolites, specifically flavonoids (Huang et al., 2006). The genus of *Scutellaria* possesses almost 300 species. Among them, *Scutellaria baicalensis* is familiar in Chinese traditional medicine and is extensively used to treat allergies, hyperlipidemia, arteriosclerosis, and inflammatory diseases (Huang, 2010). However, there is scant evidence on the pharmaceutical activities of the Iranian species, *Scutellaria pinnatifida* (*S. pinnatifida*) (Souria et al., 2004; Sauvage et al., 2010; Boozari et al., 2015).

Scutellaria pinnatifida (*S. pinnatifida*) extracts exhibited an antifibrillation activity on α-SN accumulation and neuroprotective characteristics on PC12 and primary dopaminergic neurons. *S. pinnatifida* extracts demonstrated that the extractions of dichloromethane (DCMEx) and *n*-butanol (BuOHEx) remarkably attenuated α-SN fibrillation. Studies have also shown that *S. pinnatifida* comprises a huge quantity of flavonoids with significant antioxidant activities. In addition, HPLC and mass spectroscopy studies on the DCMEx identified the baicalein in this extract (Sashourpour et al., 2017). Various studies on the mechanism of α-SN aggregation demonstrated that the toxic oligomers of α-syn and other aggregate-making proteins share similar conformation, and this includes a vast quantity of β-sheet secondary structure that augments the protein-cellular membrane interactions resulting in cytotoxicity (Roberts and Brown, 2015; Kayed et al., 2003).

Several flavonoids like baicalein and epigallocatechin gallate offer neuroprotection against misfolded-protein-stimulated toxicity, preferably α-SN (Hong et al., 2008; Lu et al., 2011). It has been found that flavonoids are robust antioxidant compounds (Shieh et al., 1999). The level of flavonoids was found to be directly proportional to the antifibrillation characteristics of the extracts. In line with this, DCMEx, the extract containing an increased quantity of flavonoids, showed significant antifibrillation effects. Conversely, HexEx with significantly low flavonoids exhibited the lowest suppressive effect on α-SN fibril formation. Therefore, it can be corroborated that the antifibrillation activity of DCMEx may be primarily due to its flavonoids, in line with earlier studies (Masuda et al., 2006; Meng et al., 2009, 2010b). It has also been shown that MeOHEx and a few subextracts exert an increased free radical scavenging activity with a robust association between the flavonoid amount and the decrease of ROS generation stimulated by α-SN. The study has also emphasized that EtOAcEx and DCMEx were robust extracts in foraging DPPH radicals. Therefore, it can be concluded that the larger the quantity of flavonoids, the higher the antioxidant activity (Bors et al., 1990; Nicholls and Budd, 2000).

Hypericum perforatum

Hypericum perforatum, an extensively examined medicinal plant popularly known as St. John's Wort, has been found to control levels of the monoamines dopamine, norepinephrine, and serotonin via attenuation of reuptake (Nathan, 2001; Singer et al., 1999). Hypericin, a bioactive compound of this plant, has been observed to attenuate the release of glutamate (Chang and Wang, 2010), and pharmacological agents that intervene with glutamate signaling were noticed to decrease symptoms and deliberate disease development in preclinical models of PD (Johnson et al., 2009). Clinical trials revealed profound protection against depressive symptoms in patients receiving an extract made of *Hypericum perforatum* versus placebo (Gaster and Holroyd, 2000; Schrader, 2000; Woelk, 2000; Jacquet et al., 2017).

Mentha × piperita (peppermint)

Mentha × piperita (peppermint) is an extensively used medicinal plant with substantial pharmacological activities. The oil isolated from peppermint comprises primarily menthol, menthone, isomenthone, and eucalyptol, despite the variation in quantities of these compounds among various extracts (McKay and Blumberg, 2006). Earlier studied characteristics of peppermint plant extracts or essential oils, which are substantially applicable to PD therapy, involve analgesic effects in mouse models of pain and the capability to eliminate free radicals (McKay and Blumberg, 2006; Riachi and De Maria, 2015).

Cyanobacteria spirulina

Cyanobacteria spirulina, consumed in some traditional cultures and currently used as a nutraceutical additive, offers strong antioxidant activity in rodent models, especially those stimulated by oxidant stress. Multiple lines of evidence have shown that spirulina is enriched with the protein phycocyanin that harbors a covalently conjugated chromophore—phycocyanobilin (PhyCB)—proficient in collecting light energy responsible for ATP generation in the organism (Romay et al., 2003; Liu et al., 2016). Certainly, there is an indication that phycocyanorubin contributes to bilirubin's efficacy in attenuating NADPH oxidase complexes (Mccarty, 2007; Zheng et al., 2013). This renders an adequate explanation for the capability of orally introduced phycocyanin, whole spirulina, or free PhyCB to show marked antioxidant activity in rodent studies. Therefore, orally administered spirulina has been observed to combat PD in three rodent models (Chamorro et al., 2006; Pabón et al., 2012; Tobon-Velasco et al., 2013; Lima et al., 2017; Mccarty, 2007).

Omega-3 fatty acids

A wealth of studies has shown that diets containing long-chain omega-3 fatty acids, usually available in fish—preferably docosahexaenoic acid (DHA), induced neuroprotection in a wide range of PD rodent models

(Agar et al., 2016; Barros et al., 2017; Mori et al., 2017; Gomez-Soler et al., 2018; Hernando et al., 2019; Chitre et al., 2020). Mitigated expression of inducible NOS or nitrite was observed in the striatum of PD rodents administered with omega-3s, indicating the inhibition of microglia iNOS initiation as a probable mechanism of protection (Barros et al., 2017; Mori et al., 2017). These results are probably clinically applicable, since enhanced dietary unsaturated fats such as long-chain omega-3s have been linked to reduced PD risk in several related epidemiological studies. A probable method is that DHA positively affects PD risk through enhanced production of the DHA-derived mediators resolvin D1 and resolvin D2, which protected the substantia nigra of rodents when administered intrathecally and that exert protection in cell culture models of PD (Krashia et al., 2019).

Conclusions and future perspectives

Therapeutic effects of drugs reduce as the disease progresses, then render merely symptomatic treatment. Therefore, innovative therapeutic strategies are needed that can stop or suppress disease progression. A plethora of plant extracts have been found to be neuroprotective, yet a few plants fail to show neuroprotection and may also be lethal; the essential oil of *Myrtus communis*, well known as Zoroastrian for its enhanced aroma (Morshedi and Nasouti, 2016), may cause α-syn fibrillation and toxicity in human neuroblastoma cells (Morshedi and Nasouti, 2016). Therefore, there is an increasing need to investigate the toxicity of plant extracts before their use in neuroprotection. In addition, substantial insight into the plant extracts and their bioactive agents is required to recommend their dietary or therapeutic use (Javed et al., 2019).

References

Agar, A., Parlak, H., Tanriover, G., Dilmac, S., Ulker, S.N., Birsen, L., et al., 2016. The protective mechanism of docosahexaenoic acid in mouse model of Parkinson: the role of hemeoxygenase. Neurochem. Int. 101, 110–119.

Ahlskog, J.E., 1994. Treatment of Parkinson's disease. From theory to practice. Postgrad. Med. 95 (52–54), 57–58.

Amro, M.S., Teoh, S.L., Norzana, A.G., Srijit, D., 2018. The potential role of herbal products in the treatment of Parkinson's disease. Clin. Ter. 169, e23–e33.

Attia, A.H., Gadelkarim, M., Morsi, M., Awad, K., Elnenny, M., Ghanem, E., et al., 2017. Meta-analysis of creatine for neuroprotection against Parkinson's disease. CNS Neurol. Disord. Drug Targets 16, 169–175.

Barros, A.S., Crispim, R.Y.G., Uchoa, J.C., Souza, R.B., Lemos, J.C., Filho, G.C., et al., 2017. Impact of the chronic omega-3 fatty acids supplementation in hemiparkinsonism model induced by 6-hydroxydopamine in rats. Basic Clin. Pharmacol. Toxicol. 120, 523–531.

Bartels, T., Choi, J.G., Selkoe, D.J., 2011. a-Synuclein occurs physiologically as a helically folded tetramer that resists aggregation. Nature 477, 107–110.

Beal, M.F., Oakes, D., Shoulson, I., Henchcliffe, C., Galpern, W.R., Haas, R., et al., 2014. Parkinson Study GroupQE3 Investigators, a randomized clinical trial of high-dosage coenzyme Q10 in early Parkinson disease: no evidence of benefit. JAMA Neurol. 71, 543–552.

Bendor, J.T., Logan, T.P., Edwards, R.H., 2013. The function of -synuclein. Neuron 79, 1044–1066.

Bergström, A.L., Kallunki, P., Fog, K., 2016. Development of passive immunotherapies for synucleinopathies. Mov. Disord. 31, 203–213.

Bernd, K., 2008. Phytochemical approach and bioanalytical strategy to develop chaperone-based medications. Open Biochem. J. 2, 44–48.

Berrocal, R., Vasudevaraju, P., Indi, S.S., Sambasiva Rao, K.R., Rao, K.S., 2014. In vitro evidence that an aqueous extract of Centella asiatica modulates a-synuclein aggregation dynamics. J. Alzheimers Dis. 39, 457–465.

Beutler, J.A., 2009. Natural products as a foundation for drug discovery. Curr. Protoc. Pharmacol. 9, 9–11.

Boozari, M., Mohammadi, A., Asili, J., Emami, S.A., Tayarani-Najaran, Z., 2015. Growth inhibition and apoptosis induction by Scutellaria pinnatifida A. Ham. on HL-60 and K562 leukemic cell lines. Environ. Toxicol. Pharmacol. 39, 307–312.

Borghi, R., Marchese, R., Negro, A., Marinelli, L., Forloni, G., Zaccheo, D., et al., 2000. Full length alpha-synuclein is present in cerebrospinal fluid from Parkinson's disease and normal subjects. Neurosci. Lett. 287, 65–67.

Bors, W., Heller, W., Michel, C., Saran, M., 1990. Radical chemistry of flavonoid antioxidants. In: Antioxidants in Therapy and Preventive Medicine. Springer, pp. 165–170.

Broersen, K., Ruiperez, V., Davletov, B., 2018. Structural and aggregation properties of alpha-synuclein linked to phospholipase A2 action. Protein Pept. Lett. 25, 368–378.

Brundin, P., Dave, K.D., Kordower, J.H., 2017. Therapeutic approaches to target alpha-synuclein pathology. Exp. Neurol. 298, 225–235.

Burre, J., Sharma, M., Tsetsenis, T., Buchman, V., Etherton, M.R., Sudhof, T.C., 2010. α-Synuclein promotes SNARE-complex assembly in vivo and in vitro. Science 329, 1663–1667.

Büttner, S., Broeskamp, F., Sommer, C., Markaki, M., Habernig, L., Alavian-Ghavanini, A., et al., 2014. Spermidine protects against α-synuclein neurotoxicity. Cell Cycle 13, 3903–3908.

Cannon, J.R., Greenamyre, J.T., 2013. Gene-environment interactions in Parkinson's disease: specific evidence in humans and mammalian models. Neurobiol. Dis. 57, 38–46.

Caruana, M., Högen, T., Levin, J., Hillmer, A., Giese, A., Vassallo, N., 2011. Inhibition and disaggregation of a-synuclein oligomers by natural polyphenolic compounds. FEBS Lett. 585, 1113–1120.

Chamorro, G., Pérez-Albiter, M., Serrano-García, N., Mares-Sámano, J.J., Rojas, P., 2006. Spirulina maxima pretreatment partially protects against 1-methyl-4-phenyl-1,2,3,6-tetrahydropyridine neurotoxicity. Nutr. Neurosci. 9, 207–212.

Chang, Y., Wang, S.J., 2010. Hypericin, the active component of St. John's wort, inhibits glutamate release in the rat cerebro cortical synaptosomes via a mitogen-activated protein kinase-dependent pathway. Eur. J. Pharmacol. 634, 53–61.

Cheon, S.M., Jang, I., Lee, M.H., Kim, D.K., Jeon, H., Cha, D.S., 2017. Sorbus alnifolia protects dopaminergic neurodegeneration in Caenorhabditis elegans. Pharm. Biol. 55, 481–486.

Chitre, N.M., Wood, B.J., Ray, A., Moniri, N.H., Murnane, K.S., 2020. Docosahexaenoic acid protects motor function and increases dopamine synthesis in a rat model of Parkinson's disease via mechanisms associated with increased protein kinase activity in the striatum. Neuropharmacology 167, 107976.

Collier, T.J., Kanaan, N.M., Kordower, J.H., 2017. Aging and Parkinson's disease: different sides of the same coin? Mov. Disord. 32, 983–990.

Cox, D., Selig, E., Griffin, M.D., Carver, J.A., Ecroyd, H., 2016. Small heat-shock proteins prevent a-synuclein aggregation via transient interactions and their efficacy is affected by the rate of aggregation. J. Biol. Chem. 291, 22618–22629.

da Costa, I.M., Cavalcanti, J.R.L.P., de Queiroz, D.B., de Azevedo, E.P., do Rêgo, A.C.M., Araújo Filho, I., et al., 2017. Supplementation with herbal extracts to promote behavioral and neuroprotective effects in experimental models of Parkinson's disease: a systematic review. Phytother. Res. 31, 959–970.

Dimant, H., Ebrahimi-Fakhari, D., McLean, P.J., 2012. Molecular chaperones and co-chaperones in Parkinson disease. Neuroscientist 18, 589–601.

Friesen, E.L., De Snoo, M.L., Rajendran, L., Kalia, L.V., Kalia, S.K., 2017. Chaperone-based therapies for disease modification in Parkinson's disease. Parkinsons Dis. 2017, 5015307.

Gadad, B.S., Subramanya, P.K., Pullabhatla, S., Shantharam, I.S., Rao, K.S., 2012. Curcumin-glucoside, a novel synthetic derivative of curcumin, inhibits a-synuclein oligomer formation: relevance to Parkinson's disease. Curr. Pharm. Des. 18, 76–84.

Gaster, B., Holroyd, J., 2000. St John's wort for depression: a systematic review. Arch. Intern. Med. 160, 152–156.

Gautam, S., Karmakar, S., Bose, A., Chowdhury, P.K., 2014. β-Cyclodextrin and curcumin, a potent cocktail for disaggregating and/or inhibiting amyloids: a case study with a-synuclein. Biochemistry 53, 4081–4083.

Gautam, S., Karmakar, S., Batra, R., Sharma, P., Pradhan, P., Singh, J., et al., 2017. Polyphenols in combination with b-cyclodextrin can inhibit and disaggregate α-synuclein amyloids under cell mimicking conditions: a promising therapeutic alternative. Biochim. Biophys. Acta 1865, 589–603.

Gelders, G., Baekelandt, V., Van der Perren, A., 2018. Linking neuroinflammation and neurodegeneration in Parkinson's disease. J. Immunol. Res. 2018, 4784268.

Ghiglieri, V., Calabrese, V., Calabresi, P., 2018. Alpha-synuclein: from early synaptic dysfunction to neurodegeneration. Front. Neurol. 9, 295.

Gomez-Soler, M., Cordobilla, B., Morato, X., Fernandez-Duenas, V., Domingo, J.C., Ciruela, F., 2018. Triglyceride form of docosahexaenoic acid mediates neuroprotection in experimental parkinsonism. Front. Neurosci. 12, 604.

Hansen, C., Li, J.Y., 2012. Beyond α-synuclein transfer: pathology propagation in Parkinson's disease. Trends Mol. Med. 18, 248–255.

Henrich, C.J., Beutler, J.A., 2013. Matching the power of high throughput screening to the chemical diversity of natural products. Nat. Prod. Rep. 30, 1284–1298.

Hernando, S., Requejo, C., Herran, E., Ruiz-Ortega, J.A., Morera-Herreras, T., Lafuente, J.V., et al., 2019. Beneficial effects of n-3 polyunsaturated fatty acids administration in a partial lesion model of Parkinson's disease: the role of glia and NRf2 regulation. Neurobiol. Dis. 121, 252–262.

Herrera, A., Muñoz, P., Steinbusch, H.W.M., Segura-Aguilar, J., 2017. Are dopamine oxidation metabolites involved in the loss of dopaminergic neurons in the nigrostriatal system in Parkinson's disease? ACS Chem. Neurosci. 8, 702–711.

Hong, D.P., Fink, A.L., Uversky, V.N., 2008. Structural characteristics of α-synuclein oligomers stabilized by the flavonoid baicalein. J. Mol. Biol. 383, 214–223.

Huang, K.C., 2010. The Pharmacology of Chinese Herbs. CRC Press.

Huang, W.H., Lee, A.R., Yang, C.H., 2006. Antioxidative and anti-inflammatory activities of polyhydroxy flavonoids of Scutellaria baicalensis GEORGI. Biosci. Biotechnol. Biochem. 70, 2371–2380.

Hussain, G., Rasul, A., Anwar, H., Aziz, N., Razzaq, A., Wei, W., et al., 2018. Role of plant derived alkaloids and their mechanism in neurodegenerative disorders. Int. J. Biol. Sci. 14, 341–357.

Inoue, E., Shimizu, Y., Masui, R., Hayakawa, T., Tsubonoya, T., Hori, S., et al., 2018. Effects of saffron and its constituents, crocin-1, crocin-2, and crocetin on a-synuclein fibrils. J. Nat. Med. 72, 274–279.

Jacquet, A., Tambe, M.A., Ma, S.Y., McCabe, G.P., Vest, J.H.C., Rochet, J., 2017. Pikuni-Blackfeet traditional medicine: neuroprotective activities of medicinal plants used to treat Pakinson's disease-related symptoms. J. Ethnopharmacol. 206, 393–407.

Jadiya, P., Khan, A., Sammi, S.R., Kaur, S., Mir, S.S., Nazira, A., et al., 2011. Anti-parkinsonian effects of Bacopa monnieri: insights from transgenic and pharmacological Caenorhabditis elegans models of Parkinson's disease. Biochem. Biophys. Res. Commun. 413, 605–610.

Javed, H., Nagoor Meeran, M.F., Azimullah, S., Adem, A., Sadek, B., Ojha, S.K., 2019. Plant extracts and phytochemicals targeting α-synuclein aggregation in Parkinson's disease models. Front. Pharmacol. 9, 1555.

Jha, N.N., Ghosh, D., Das, S., Anoop, A., Jacob, R.S., Singh, P.K., et al., 2016. Effect of curcumin analogs on a-synuclein aggregation and cytotoxicity. Sci. Rep. 6, 28511.

Ji, H.F., Shen, L., 2014. The multiple pharmaceutical potential of curcumin in Parkinson's disease. CNS Neurol. Disord. Drug Targets 13, 369–373.

Jiang, M., Porat-Shliom, Y., Pei, Z., Cheng, Y., Xiang, L., Sommers, K., 2010. Baicalein reduces E46K alpha-synuclein aggregation in vitro and protects cells against E46K alpha-synuclein toxicity in cell models of familiar Parkinsonism. J. Neurochem. 114, 419–429.

Johnson, K.A., Conn, P.J., Niswender, C.M., 2009. Glutamate receptors as therapeutic targets for Parkinson's disease. CNS Neurol. Disord. Drug Targets 8, 475–491.

Kalia, L.V., Kalia, S.K., Lang, A.E., 2015. Disease-modifying strategies for Parkinson's disease. Mov. Disord. 30, 1442–1450.

Kastenholz, B., Garfin, D.E., 2009. Medicinal plants: a natural chaperones source for treating neurological disorders. Protein Pept. Lett. 16, 116–120.

Kayed, R., Head, E., Thompson, J.L., McIntire, T.M., Milton, S.C., et al., 2003. Common structure of soluble amyloid oligomers implies common mechanism of pathogenesis. Science 300, 486–489.

Krashia, P., Cordella, A., Nobili, A., La Barbera, L., Federici, M., Leuti, A., et al., 2019. Blunting neuroinflammation with resolvin D1 prevents early pathology in a rat model of Parkinson's disease. Nat. Commun. 10, 3945.

Kundu, P., Das, M., Tripathy, K., Sahoo, S.K., 2016. Delivery of dual drug loaded lipid based nanoparticles across the blood-brain barrier imparts enhanced neuroprotection in a rotenone induced mouse model of Parkinson's disease. ACS Chem. Nerosci. 7, 1658–1670.

Lashuel, H.A., Overk, C.R., Oueslati, A., Masliah, E., 2013. The many faces of α-synuclein: from structure and toxicity to therapeutic target. Nat. Rev. Neurosci. 14, 38–48.

Lauterbach, E.C., Victoroff, J., Coburn, K.L., Shillcutt, S.D., Doonan, S.M., Mendez, M.F., 2010. Psychopharmacological neuroprotection in neurodegenerative disease: assessing the preclinical data. J. Neuropsychiatry Clin. Neurosci. 22, 8–18.

Lee, J.H., Lee, I.H., Choe, Y.J., Kang, S., Kim, H.Y., Gai, W.P., et al., 2009. Real-time analysis of amyloid fibril formation of alpha-synuclein using a fibrillation-state-specific fluorescent probe of JC-1. Biochem. J. 418, 311–323.

Lester, H.A., Xiao, C., Srinivasan, R., Son, C.D., Miwa, J., Pantoja, R., et al., 2009. Nicotine is a selective pharmacological chaperone of acetylcholine receptor number and stoichiometry. Implications for drug discovery. AAPS J. 11, 167–177.

Lima, F.A.V., Joventino, I.P., Neves, K., Leal, L.K.A.M., De Almeida, A.C., Carmo, M., et al., 2017. Neuroprotective activities of Spirulina platensis in the 6-OHDA model of Parkinson's disease are related to its anti-inflammatory effects. Neurochem. Res. 42, 3390–3400.

Liu, Q., Huang, Y., Zhang, R., Cai, T., Cai, Y., 2016. Medical application of Spirulina platensis derived C-phycocyanin. Evid. Based Complement. Alternat. Med. 2016, 7803846.

Lu, J.H., Ardah, M.T., Durairajan, S.S.K., Liu, L.F., Xie, L.X., Fong, W.D., et al., 2011. Baicalein inhibits formation of α-synuclein oligomers within living cells and prevents Aβ peptide fibrillation and oligomerisation. Chembiochem 12, 615–624.

Marchiani, A., Mammi, S., Siligardi, G., Hussain, R., Tessari, I., Bubacco, L., et al., 2013. Small molecules interacting with α-synuclein: antiaggregating and cytoprotective properties. Amino Acids 45, 327–338.

Masuda, M., Suzuki, N., Taniguchi, S., Oikawa, T., Nonaka, T., Iwatsubo, T., et al., 2006. Small molecule inhibitors of α-synuclein filament assembly. Biochemistry 45, 6085–6094.

Mazo, N.A., Echeverria, V., Cabezas, R., Avila-Rodriguez, M., Tarasov, V.V., Yarla, N.S., et al., 2017. Medicinal plants as protective strategies against Parkinson's disease. Curr. Pharm. Des. 23, 4180–4188.

Mccarty, M.F., 2007. Clinical potential of Spirulina as a source of phycocyanobilin. J. Med. Food 10, 566–570.

McKay, D.L., Blumberg, J.B., 2006. A review of the bioactivity and potential health benefits of peppermint tea (Mentha piperita L.). Phytother. Res. 20, 619–633.

Meng, X., Munishkina, L.A., Fink, A.L., Uversky, V.N., 2009. Molecular mechanisms underlying the flavonoid-induced inhibition of alpha-synuclein fibrillation. Biochemistry 48, 8206–8224.

Meng, F., Abedini, A., Plesner, A., Verchere, C.B., Raleigh, D.P., 2010a. The flavanol (−)-epigallocatechin 3-gallate inhibits amyloid formation by islet amyloid polypeptide, disaggregates amyloid fibrils, and protects cultured cells against IAPP-induced toxicity. Biochemistry 49, 8127–8133.

Meng, X., Munishkina, L.A., Fink, A.L., Uversky, V.N., 2010b. Effects of various flavonoids on the Parkinson's disease., p. 2010 (650794).

Molino, S., Dossena, M., Buonocore, D., Ferrari, F., Venturini, L., Ricevuti, G., et al., 2016. Polyphenols in dementia: from molecular basis to clinical trials. Life Sci. 61, 69–77.

Moore, D.J., West, A.B., Dawson, V.L., Dawson, T.M., 2005. Molecular pathophysiology of Parkinson's disease. Annu. Rev. Neurosci. 28, 57–87.

Morgan, L.A., Grundmann, O., 2017. Preclinical and potential applications of common western herbal supplements as complementary treatment in Parkinson's disease. J. Diet Suppl. 14, 453–466.

Mori, M.A., Delattre, A.M., Carabelli, B., Pudell, C., Bortolanza, M., Staziaki, P.V., et al., 2017. Neuroprotective effect of omega-3 polyunsaturated fatty acids in the 6-OHDA model of Parkinson's disease is mediated by a reduction of inducible nitric oxide synthase. Nutr. Neurosci. 21, 341–351.

Morshedi, D., Nasouti, M., 2016. Essential oils may lead α-synuclein towards toxic fibrils formation. Parkinsons. Dis. 2016, 6219249.

Nathan, P.J., 2001. Hypericum perforatum (St John's Wort): a non-selective reuptake inhibitor? A review of the recent advances in its pharmacology. J. Psychopharmacol. 15, 47–54.

Nicholls, D.G., Budd, S.L., 2000. Mitochondria and neuronal survival. Physiol. Rev. 80, 315–360.

Obulesu, M., 2021. Turmeric and Curcumin in Neurodegenerative Diseases, first ed. Elsevier.

Ono, K., Yamada, M., 2006. Antioxidant compounds have potent antifibrillogenic and fibril-destabilizing effects for alpha-synuclein fibrils in vitro. J. Neurochem. 97, 105–115.

Pabón, M.M., Jernberg, J.N., Morganti, J., Contreras, J., Hudson, C.E., Klein, R.L., et al., 2012. A spirulina-enhanced diet provides neuroprotection in an α-synuclein model of Parkinson's disease. PLoS One 7, e45256.

Pandey, N., Strider, J., Nolan, W.C., Yan, S.X., Galvin, J.E., 2008. Curcumin inhibits aggregation of alpha-synuclein. Acta Neuropathol. 115, 479–489.

Riachi, L.G., De Maria, C.A., 2015. Peppermint antioxidants revisited. Food Chem. 176, 72–81.

Ringe, D., Petsko, G.A., 2009. What are pharmacological chaperones and why are they interesting? J. Biol. 8, 80.

Roberts, H.L., Brown, D.R., 2015. Seeking a mechanism for the toxicity of oligomeric α-synuclein. Biomol. Ther. 5, 282–305.

Romay, C., Gonzalez, R., Ledon, N., Remirez, D., Rimbau, V.C., 2003. Phycocyanin: a biliprotein with antioxidant, anti-inflammatory and neuroprotective effects. Curr. Protein Pept. Sci. 4, 207–216.

Ruipérez, V., Darios, F., Davletov, B., 2010. Alpha-synuclein, lipids and Parkinson's disease. Prog. Lipid Res. 49, 420–428.

Sarrafchi, A., Bahmani, M., Shirzad, H., Rafieian-Kopaei, M., 2016. Oxidative stress and Parkinson's disease: new hopes in treatment with herbal antioxidants. Curr. Pharm. Des. 22, 238–246.

Sashourpour, M., Zahri, S., Radjabian, T., Ruf, V., Pan-Montojo, F., Morshedi, D., 2017. A study on the modulation of alpha-synuclein fibrillation by Scutellaria pinnatifida extracts and its neuroprotective properties. PLoS One 12, e0184483.

Sauvage, S., Samson, E., Granger, M., Majumdar, A., Nigam, P., Naharz, L., et al., 2010. Assessment of free-radical scavenging and antibacterial activities, and brine shrimp toxicity of Scutellaria pinnatifida. Orient Pharm. Exp. Med. 10, 304–309.

Schrader, E., 2000. Equivalence of St John's wort extract (Ze 117) and fluoxetine: a randomized, controlled study in mild-moderate depression. Int. Clin. Psychopharmacol. 15, 61–68.

Shaltiel-Karyo, R., Davidi, D., Frenkel-Pinter, M., Ovadia, M., Segal, D., Gazit, E., 2012. Differential inhibition of α-synuclein oligomeric and fibrillar assembly in Parkinson's disease model by cinnamon extract. Biochim. Biophys. Acta 1820, 1628–1635.

Shieh, D., Liu, L.T., Lin, C.C., 1999. Antioxidant and free radical scavenging effects of baicalein, baicalin and wogonin. Anticancer Res. 20, 2861–2865.

Shoval, H., Weiner, L., Gazit, E., Levy, M., Pinchuk, I., Lichtenberg, D., 2008. Polyphenol-induced dissociation of various amyloid fibrils results in a methionine-independent formation of ROS. Biochim. Biophys. Acta 1784, 1570–1577.

Siddique, Y.H., Mujtaba, S.F., Jyoti, S., Naz, F., 2013. GC-MS analysis of Eucalyptus citriodora leaf extract and its role on the dietary supplementation in transgenic Drosophila model of Parkinson's disease. Food Chem. Toxicol. 55, 29–35.

Singer, A., Wonnemann, M., Muller, W.E., 1999. Hyperforin, a major antidepressant constituent of St. John's Wort, inhibits serotonin uptake by elevating free intracellular Na+1. J. Pharmacol. Exp. Ther. 290, 1363–1368.

Souria, E., Amin, G., Dehmobed, S., Nazifia, A., Farsama, H., 2004. Antioxidative activity of sixty plants from Iran. Iran J. Pharm. Res. 3, 55–59.

Srinivasan, R., Henderson, B.J., Lester, H.A., Richards, C.I., 2014. Pharmacological chaperoning of nAChRs: a therapeutic target for Parkinson's disease. Pharmacol. Res. 83, 20–29.

Taebnia, N., Morshedi, D., Yaghmaei, S., Aliakbari, F., Rahimi, F., Arpanaei, A., 2016. Curcumin-loaded amine-functionalized mesoporous silica nanoparticles inhibit a-synuclein fibrillation and reduce its cytotoxicity-associated effects. Langmuir 32, 13394–13402.

Tobon-Velasco, J.C., Palafox-Sánchez, V., Mendieta, L., Garcia, E., Santamaría, A., Cevallos, G.C., et al., 2013. Antioxidant effect of Spirulina (Arthrospira) maxima in a neurotoxic model caused by 6-OHDA in the rat striatum. J. Neural Transm. 120, 1179–1189.

Tompa, P., 2005. The interplay between structure and function in intrinsically unstructured proteins. FEBS Lett. 579, 3346–3354.

Török, N., Majláth, Z., Szalárdy, L., Vécsei, L., 2016. Investigational α-synuclein aggregation inhibitors: hope for Parkinson's disease. Expert Opin. Investig. Drugs 25, 1281–1294.

Ur Rasheed, M.S., Tripathi, M.K., Mishra, A.K., Shukla, S., Singh, M.P., 2016. Resveratrol protects from toxin-induced Parkinsonism: a plethora of proofs hitherto petty translational value. Mol. Neurobiol. 53, 2751–2760.

Uversky, V.N., Dunker, A.K., 2010. Understanding protein non-folding. Biochim. Biophys. Acta 1804, 1231–1264.

Wang, Z.Y., Liu, J.Y., Yang, C.B., Malampati, S., Huang, Y.Y., Li, M.X., et al., 2017. Neuroprotective natural products for the treatment of Parkinson's disease by targeting the autophagy-lysosome pathway: a systematic review. Phytother. Res. 31, 1119–1127.

Woelk, H., 2000. Comparison of St John's wort and imipramine for treating depression: randomised controlled trial. BMJ 321, 536–539.

Zhang, H., Bai, L., He, J., Zhong, L., Duan, X., Ouyang, L., et al., 2017. Recent advances in discovery and development of natural products as source for anti-Parkinson's disease lead compounds. Eur. J. Med. Chem. 141, 257–272.

Zheng, J., Inoguchi, T., Sasaki, S., Maeda, Y., Mccarty, M.F., Fujii, M., et al., 2013. Phycocyanin and phycocyanobilin from Spirulina platensis protect against diabetic nephropathy by inhibiting oxidative stress. Am. J. Physiol. Integr. Comp. Physiol. 304, R110–R120.

Zhu, M., Rajamani, S., Kaylor, J., Han, S., Zhou, F., Fink, A.L., 2004. The flavonoid baicalein inhibits fibrillation of alpha-synuclein and disaggregates existing fibrils. J. Biol. Chem. 279, 26846–26857.

CHAPTER 3

Panoply of plant extracts in the treatment of prion diseases

Introduction

Pathology

Numerous neurodegenerative diseases (NDs) like Alzheimer's disease (AD), Parkinson's disease, Huntington's disease (HD), amyotrophic lateral sclerosis (ALS), and frontotemporal dementia are collectively grouped under prion-like diseases since they mimic a few pathological events observed in prion disease (PD) (Toni et al., 2017). PD are a class of infectious diseases usually initiated by neurodegeneration induced by misfolded and accumulated proteins such as PrPsc in specific mammalian species like humans (Prasad and Bondy, 2018). The discovery of PD symptoms such as scrapie in sheep and goats dates back to 1730s. A contagious neurological disease known as kuru, akin to Creutzfeldt-Jakob disease (CJD), was found in the Fore tribe of Papua, New Guinea (Gajdusek and Zigas, 1957; Prasad and Bondy, 2018). Dr. Stanley Prusiner isolated infective scrapie, which provoked neurodegeneration in the brains of sheep and goats, and coined the term "prion" (Prusiner, 1982). Despite the consideration of NDs like AD as PDs in 2012, β-sheets of AD were found noninfective unlike sheets of PrPsc that provoke transmissible spongiform encephalopathy (TSE) (Prusiner, 2012; Prasad and Bondy, 2018). The proteins primarily playing a role in prion-like diseases, such as Aβ for AD, tau for tauopathies, and α-synuclein for synucleinopathies, also act as prion proteins in seeding mechanisms (Yamamoto et al., 2005; Nonaka et al., 2010), spreading (Frost et al., 2009; Hansen et al., 2011), and cell interaction methods (Yamamoto et al., 2005; Nonaka et al., 2010; Toni et al., 2017).

PDs perturb the brain and the nervous system, initiating the decline of mental and physical activities, amyloid plaque formation, neuronal death, and generation of holes in the cortex that appears like a sponge when observed under a microscope. The usual clinical signs observed in humans probably involve memory issues and personality alterations, depression, loss of coordination, involuntary movements, or insomnia.

Plant Extracts in Neurodegenerative Diseases
https://doi.org/10.1016/B978-0-323-95762-5.00002-3

The regular observations in the advanced stages of the disease include dementia and an inability to move or speak (Aguzzi et al., 2007; Sanchez et al., 2017). Despite the appropriate physiologic role of prion proteins as a cellular signaling molecule when the onset of infectious proteins occurs, all the prion proteins transform into a pathological form (Shi et al., 2016; Slapsak et al., 2016; Kim et al., 2017; Sanchez et al., 2017). Cellular prion protein has been known to aggregate with amyloid β (Aβ) oligomers with a high affinity, which play a key role in Alzheimer's pathology (Sanchez et al., 2017).

Numerous studies demonstrated that infectious prions are assimilated in the environment and soil by feces, saliva, blood, urine, or placenta through infected animals and decaying carcasses (Mathiason et al., 2006; Haley et al., 2009; Tamguney et al., 2009; Maddison et al., 2010; Haley et al., 2011; Terry et al., 2011). Studies corroborated that infectious prions accumulate in soil and induce infections for a long time, thus spreading TSE (Johnson et al., 2006; Seidel et al., 2007; Johnson et al., 2007). PrPsc bound to leaves and roots of wheat grass plants showed an ability to seed prion replication in vitro. Unexpectedly, the minute quantity of PrPsc excreted in urine and feces from infected hamster or cervids profoundly contaminated plant tissue. Higher quantities of excreted PrPsc have been retained by plants, leaves, and roots. In addition, leaves can be infected even by spraying them with something containing a prion. Prion-contaminated plants spread PD to animals after ingestion, generating a 100% attack rate. Despite the ambiguity of plants' role in absorption of soil proteins and transport to aerial parts, a few studies confirmed this phenomenon (McLaren et al., 1960; Jensen and McLaren, 1960; Paungfoo-Lonhienne et al., 2008). The prions' profound resistance to degradation and their efficacy to span biological barriers probably plays a crucial role in this process. Pastures were found to contain prions for 2 years following exposure (Miller et al., 2004). The mechanism of prion and plant binding, retention, uptake, and transport is elusive.

A wealth of studies has shown that enhanced oxidative stress (Brazier et al., 2006; Chen et al., 2008) and inflammation (Crespo et al., 2012; Algarzae et al., 2012; Carroll et al., 2015) promote the progression of PD (Prasad and Bondy, 2018). Mounting evidence has shown that oxidative stress plays a major role in the pathogenesis of PD, and antioxidants were found to curtail neurotoxicity in both cell cultures and animal models (Brazier et al., 2008; Mizrahi et al., 2014; Asuni et al., 2015; Lee et al., 2015; Prasad and Bondy, 2018).

Plants and prions

Mounting evidence has shown that infectious prion proteins bind to plants. Minute quantities of PrPsc available in brain homogenate or excretory products like urine and feces transfer to wheat grass roots and leaves. When these leaves are taken by wild-type hamsters, infection is observed in them. In addition, leaves sprayed with prion-containing preparation accumulated PrPsc for prolonged duration in plants. PDs are communicable animal-to-animal, animal-to-human, and human-to-human. Nevertheless, complete mechanism, agents, and biological processes that regulate the transmission of this specific infectious agent are elusive. The transmission of a few forms of TSEs like CJD, kuru, and BSE occurs due to the intake of affected meat and meat products (Collinge, 1999; Prusiner, 2001). On the contrary, a few horizontally transmissible animal TSEs such as scrapie and CWD contaminated the environment with prions (Mathiason et al., 2009; Gough and Maddison, 2010; Bartelt-Hunt and Bartz, 2013).

Metal dyshomeostasis

Metal ions play a pivotal role as cofactors in a plethora of enzymes. However, they may also be deleterious to the cell by involving redox reactions and formation of reactive oxygen species (ROS) (Toni et al., 2017). Many studies emphasize environmental metal exposure as a contributing factor for prion and prion-like diseases such as AD and PD (Toni et al., 2017). A diet with ample metal ions and chelating agents like polyphenols can combat NDs. Epidemiological studies showed that a Mediterranean diet with fresh vegetables and low meat substantially reduces the risk of prion and prion-like diseases (Toni et al., 2017).

Epidemiological studies reveal that occupational exposure to environmental metals like copper, manganese, iron, lead, zinc, mercury, and aluminum play a key role in the pathogenesis of neurodegenerative (Obulesu et al., 2011) and prion-like diseases (Zayed et al., 1990; Rybicki et al., 1993; Gorell et al., 1997; Benedetto et al., 2009; Fukushima et al., 2010; Cannon and Greenamyre, 2011; Toni et al., 2017).

Manifold metal ions were found to bind with varying affinities to the PrPc, perturbing its conformation and ability to aggregate and impact PrPC/PrPSc alteration. The octapeptide repeats (OR) moiety of the PrPC is the primary binding site that can communicate with Cu^{2+}, Zn^{2+}, and Mn^{2+} (Jackson et al., 2001; Walter et al., 2007; Brown, 2011) thus offering enhanced stability to the protein (Benetti et al., 2014; Toni et al., 2017). PrPC has been linked to metal-based enzymatic functions (Brown et al., 1999; Rachidi et al., 2003) following the augmentation of cell antioxidant capacity (White et al., 1999; Mitteregger et al., 2007; Peggion et al., 2016). A Cu^{2+}-based neuroprotective role of PrPC by regulating *N*-methyl-D-aspartate (NMDA) receptor-nitrosylation has been recently discovered (Gasperini et al., 2015).

Plants

Bacoside-A, a natural triterpene glycoside with trisaccharide moiety, has been found to exert antiamyloid properties (Malishev et al., 2016; Sanchez et al., 2017). *Bacopa monniera*, which harbors amphiphilic compounds structurally dependent on sterol glycosides, has been extensively employed in traditional Indian medicine to combat nervous disorders and improve memory (Bammidi et al., 2001; Limpeanchob et al., 2008; Sanchez et al., 2017). Several lines of evidence has shown that bacoside-A might combat amyloid diseases like AD (Holcomb et al., 2006; Apetz et al., 2014; Sanchez et al., 2017). Studies of Malishev et al. on the interactions between bacoside-A and the 21-residue amyloidogenic element of the prion protein PrP106–126 by spectroscopy and microscopy techniques have shown that PrP106–126 provokes fibril formation at the lipid bilayer surface (Tagliavini et al., 1993; Miura et al., 2007; Sanchez et al., 2017).

Polyphenols

A plethora of polyphenols exert exemplary medicinal properties. They have been found to show antioxidant (Rice-Evans et al., 1996; Pignatelli et al., 2006; Pandey and Rizvi, 2009; Kim et al., 2014; Perron and Brumaghim, 2009), antiinflammatory (Sergent et al., 2010; Gonzalez et al., 2011), and antiamyloidogenic effects (Ono et al., 2003; Porzoor et al., 2015; Viviane et al., 2015; Sanchez et al., 2017). They are natural in origin and show appreciable therapeutic effects against AD, dementia, multiple sclerosis, ischemic stroke, Parkinson's disease, and Huntington's disease (Perez-Hernandez et al., 2016; Sanchez et al., 2017). Rauter et al. (2007) showed the substantial acetylcholinesterase (AChE) inhibitory efficacy of *Salvia sclareoides* extract, a plant usually found in Iberia (Martins et al., 2009; Sanchez et al., 2017). Additionally, plant extracts also prolonged prion propagation in a cell-dependent screening assay, and prion-binding characteristics were examined by NMR experiments (Rauter et al., 2012; Sanchez et al., 2017). A few polyphenols, being water soluble, fail to ferry the blood-brain barrier (BBB) but may act against TSE (Sanchez et al., 2017). Nevertheless, radiolabeled epigallocatechin gallate (EGCG) has been identified in mouse brains following oral administration (Suganuma et al., 1998; Sanchez et al., 2017). Some natural polyphenols that are structurally dependent on stilbenes or cinnamic acid derivatives, such as resveratrol and curcumin, exhibited antiprion activity.

Resveratrol

Resveratrol [systematic name:(*E*)-5-(4-hydroxystyryl)benzene-1,3-diol] is generated in grapes, mulberries, and nuts when invaded by pathogens (Walle, 2011). Growing lines of evidence have shown neuroprotective (Obulesu and Dendi, 2020), cardioprotective, hepatoprotective, antiinflammatory, and cancer-preventive effects (Bhat et al., 2001; Sanchez et al., 2017). Its neuroprotective activity has been observed in AD, Parkinson's disease, and PD (Sanchez et al., 2017). Studies have shown that resveratrol curbs PrP106–126, instigating neuronal cell death by promoting autophagy neuronal cells (SH-SY5Y cells and SK-N-SH cell lines) (Jeong et al., 2012). Wang et al. (2015) reported that resveratrol substantially curtailed the levels of PrPSc when SMB-S15 cells were treated with Sirt1 activators. In line with this, Wang et al. (2016) evaluated the inhibitory efficacy of resveratrol on prion aggregation in vitro and prion infectivity in vivo in scrapie-infected cell line SMB-S15. They showed that the quantity of PrPSc in SMB-S15 cells reduced in an adose-dependent manner in the presence of resveratrol with $EC50 = 0.61\ \mu M$ (Sanchez et al., 2017).

Curcumin

Curcumin [systematic name: 1,7-bis(4-hydroxy-3-methoxyphenyl)hepta-1,6-diene-3,5-dione], the vital pigment obtained from *Curcuma longa*, has been found to exert robust therapeutic efficacy in AD and Parkinson's disease (Obulesu, 2021). It has also been found to attenuate prion fibril formation (Caughey et al., 2003; Hafner-Bratkovic et al., 2008) in PD by spanning the BBB (Yang et al., 2005; Sanchez et al., 2017). Accordingly, studies have also shown the antiamyloidogenic and antioxidant efficacy of curcumin on the recombinant murine PrP in a cell-free milieu and in murine neuroblastoma (N2a) cells (Lin et al., 2013). They reported that amyloid formation was significantly lower and the length of mPrP fibrils was shorter in the presence of curcumin (Sanchez et al., 2017). N2a cells pretreated with curcumin showed 33% augmented cell viability. It has been found that fibril-provoked apoptosis was debilitated when pretreated with curcumin. Additionally, antioxidant activity of curcumin was examined against ROS levels induced by mPrP fibrils in N2a cells; it was noticed that mPrP fibril-instigated ROS was substantially scavenged by treatment with 2.5 μM of curcumin (Lin et al., 2013; Sanchez et al., 2017).

Disturbing the cell membrane without binding to specific proteins may be the reason for their numerous functions, and there is scant understanding about their binding to several effector proteins (Tachibana et al., 2004; Ingolfsson et al., 2014; Sanchez et al., 2017). Ingolfsson et al. (2014) reported that EGCG, resveratrol, and curcumin change bilayer properties and alter membrane protein function. Curcumin exhibits Pan Assay Interference Compounds (PAINS)-type behavior (Ingolfsson et al., 2014; Schneider et al., 2015). It has catechol units that are identifiable as PAINS motifs intervening with a wide range of biological events by various mechanisms (Baell, 2016). Similarly, resveratrol is oxidized forming reactive quinones that are also PAINS motifs (Zhang et al., 2009).

Polyphenolic metal chelators

Polyphenols play a pivotal role as metal chelators with the capacity to bind and chelate numerous bivalent metals like Cu^{2+}, Zn^{2+}, and Fe^{2+} (Singh et al., 2008; Mandel et al., 2011; Toni et al., 2017). Polyphenols also reduce metal absorption in the small intestine with their chelating efficacy (Landete, 2012). PAQUID cohort study conducted on humans has corroborated that polyphenols (flavonoids) curtail the risk of dementia by ~50% (Commenges et al., 2000; Toni et al., 2017).

Mediterranean diet

Commonly, a Mediterranean diet (MeD) comprises plant foods galore, a reasonable amount of fish and wine, low meat and dairy products, and olive oil as the primary fat source (Toni et al., 2017). Following a MeD curtails the risk of AD, Parkinson's disease, dementia, and cognitive decline (Scarmeas et al., 2006, 2009; Feart et al., 2009; Singh et al., 2014), despite the lack of epidemiological evidence for sporadic Creutzfeldt-Jacob disease (CJD) (Toni et al., 2017). MeD possesses several single constituents that exert neuroprotective effects (Jacobs et al., 2009; Gotsis et al., 2015). Plant polyphenols, being the primary constituents of this diet with fruits, vegetables, and wine (Manach et al., 2004), exhibit potential health benefits and cognitive functions (Baur and Sinclair, 2006; Witte et al., 2014). Plant polyphenols belong to the class of secondary metabolites available in herbal food and beverages and contain higher than 8000 phenolic structures (Tsao, 2010).

Lichens

The prion protein does not undergo degradation, and it has been a herculean task to accomplish its degradation. Interestingly, studies have shown that acetone extracts of lichens such as *Parmelia sulcata*, *Cladonia rangiferina*, and *Lobaria pulmonaria* substantially degraded prion protein in TSE-infected hamsters, mice, and deer (Johnson et al., 2011). Three lichen extracts exhibited serine protease activity with the ability to mitigate PrP levels in infected brain homogenates in vitro. Lichen tissue by itself or a water extract of the lichen tissue potentially degraded PrP. Based on these findings, the quest for several protease studies that can combat prions have been started (Langeveld et al., 2003; Hui et al., 2004; Lawson et al., 2007; Yoshioka et al., 2007; Pilon et al., 2009; Dickinson et al., 2009; Johnson et al., 2011).

Conclusions and future perspectives

There is scant knowledge about the appropriate mechanism by which pathological prion proteins accumulate and induce neurotoxicity (Sanchez et al., 2017). There is a growing need currently to develop multifarious antiprion molecules (Sanchez et al., 2017). Prion protein reduction also hastens Aβ formation, and its role in aging is probably correlated to the mechanisms entailed in neurodegeneration, thus leading to AD. Nevertheless, the role of PrPC in AD is elusive (Gasperini and Legname, 2014; Sanchez et al., 2017). Interestingly, prions provoke amyloid formation, and the quest for antiamyloidogenic compounds to combat AD probably show therapeutic efficacy against PDs.

Based on a plethora of research studies, it can be concluded that, despite the robust therapeutic efficacy of polyphenols against various diseases, their biological activity is incompletely understood. They alter numerous biological pathways and change the functions of a wide range of proteins such as membrane proteins (Chacko et al., 2010; Khan et al., 2006; Ramassamy,

2006; Sanchez et al., 2017). Since the understanding of PrPC as a binding companion of Aβ oligomers that play a pivotal role in neuronal cell death in AD, a few novel avenues in treatment have been opened (Lauren et al., 2009; Sanchez et al., 2017). Despite the efficacy of a few polyphenols to ferry the BBB, a single polyphenol may not reach the required concentrations in brain with which it can show successful metal chelation or free radical scavenging activities. However, MeD with its multifarious polyphenols induces substantial neuroprotective efficacy (Alcalay et al., 2012; Pelletier et al., 2015). It has been understood that a few nutritional factors can alter the mechanisms of neuronal dysfunction provoked by prion and prion-like proteins and the MeD can show neuroprotection against these diseases (Toni et al., 2017). Eventually, there is indeed a growing need to explore more appropriate therapeutic molecules, preferably from plant sources.

References

Aguzzi, A., Heikenwalder, M., Polumenidou, M., 2007. Insights into prion strains and neurotoxicity. Nat. Rev. Mol. Cell Biol. 8, 552–561.

Alcalay, R.N., Gu, Y., Mejia-Santana, H., Cote, L., Marder, K.S., Scarmeas, N., 2012. The association between Mediterranean diet adherence and Parkinson's disease. Mov. Disord. 27, 771–774.

Algarzae, N., Hebron, M., Miessau, M., Moussa, C.E.H., 2012. Parkin prevents cortical atrophy and Abeta-induced alterations of brain metabolism: ^{13}CNMR and magnetic resonance imaging studies in AD models. Neuroscience 225, 22–34.

Apetz, N., Munch, G., Govindaraghavan, S., Gyengesi, E., 2014. Natural compounds and plant extracts as therapeutics against chronic inflammation in Alzheimer's disease—a translational perspective, CNS and neurol dis, drug targets (formerly current drug targets-CNS & neurological disorders). CNS Neurol. Disord. Drug Targets 2014 (13), 1175–1191.

Asuni, A.A., Guridi, M., Sanchez, S., Sadowski, M.J., 2015. Antioxidant peroxiredoxin 6 protein rescues toxicity due to oxidative stress and cellular hypoxia *in vitro*, and attenuates prion-related pathology *in vivo*. Neurochem. Int. 90, 152–165.

Baell, J.B., 2016. Feeling nature's PAINS: natural products, natural product drugs and Pan Assay Interference Compounds (PAINS). J. Nat. Prod. 79, 616–628.

Bammidi, S.R., Volluri, S.S., Chippada, S.C., Avanigadda, S., Vangalapati, M.A., 2001. Review on pharmacological studies of Bacopa monniera. J. Chem. Biol. Phys. Sci. 1, 250.

Bartelt-Hunt, S.L., Bartz, J.C., 2013. Behavior of prions in the environment: implications for prion biology. PLoS Pathog. 9, e1003113.

Baur, J.A., Sinclair, D.A., 2006. Therapeutic potential of resveratrol: the *in vivo* evidence. Nat. Rev. Drug Discov. 5, 493–506.

Benedetto, A., Au, C., Aschner, M., 2009. Manganese-induced dopaminergic neurodegeneration: insights into mechanisms and genetics shared with Parkinson's disease. Chem. Rev. 109, 4862–4884.

Benetti, F., Biarnés, X., Attanasio, F., Giachin, G., Rizzarelli, E., Legname, G., 2014. Structural determinants in prion protein folding and stability. J. Mol. Biol. 426, 3796–3810.

Bhat, K.P.L., Kosmeder, J.W., Pezzuto, J.M., 2001. Biological effects of resveratrol. Antioxid. Redox Signal. 3, 1041–1064.

Brazier, M.W., Lewis, V., Ciccotosto, G.D., Klug, G.M., Lawson, V.A., Cappai, R., et al., 2006. Correlative studies support lipid peroxidation is linked to PrP(res) propagation as an early primary pathogenic event in prion disease. Brain Res. Bull. 68, 346–354.

Brazier, M.W., Doctrow, S.R., Masters, C.L., Collins, S.J., 2008. A manganese superoxide dismutase/catalase mimetic extends survival in a mouse model of human prion disease. Free Radic. Biol. Med. 45, 184–192.

Brown, D.R., 2011. Prions and manganese: a maddening beast. Metallomics 3, 229–238.

Brown, D.R., Wong, B.S., Hafiz, F., Clive, C., Haswell, S.J., Jones, I.M., 1999. Normal prion protein has an activity like that of superoxide dismutase. Biochem. J. 344, 1–5.

Cannon, J.R., Greenamyre, J.T., 2011. The role of environmental exposures in neurodegeneration and neurodegenerative diseases. Toxicol. Sci. 124, 225–250.

Carroll, J.A., Striebel, J.F., Race, B., Philips, K., Chesebro, B., 2015. Prion infection of mouse brain reveals multiple new upregulated genes involved in neuroinflammation or signal transduction. J. Virol. 89, 2388–2404.

Caughey, B., Raymond, L.D., Raymond, G.J., Maxson, L., Silveir, J., Baron, G.S., 2003. Inhibition of protease-resistant prion protein accumulation in vitro by curcumin. J. Virol. 77, 5499–5502.

Chacko, S.M., Thambi, P.T., Kuttan, R., Nishigaki, I., 2010. Beneficial effects of green tea: a literature review. Chin. Med. 5, 13.

Chen, Y.R., Yi, F.F., Li, X.Y., Wang, C.Y., Chen, L., Yang, X., et al., 2008. Resveratrol attenuates ventricular arrhythmias and improves the long-term survival in rats with myocardial infarction. Cardiovasc. Drugs Ther. 22, 479–485.

Collinge, J., 1999. Variant Creutzfeldt-Jakob disease. Lancet 354, 317–323.

Commenges, D., Scotet, V., Renaud, S., Jacqmin-Gadda, H., Barberger-Gateau, P., Dartigues, J.F., 2000. Intake of flavonoids and risk of dementia. Eur. J. Epidemiol. 16, 357–363.

Crespo, I., Roomp, K., Jurkowski, W., Kitano, H., del Sol, A., 2012. Gene regulatory network analysis supports inflammation as a key neurodegeneration process in prion disease. BMC Syst. Biol. 6, 132.

Dickinson, J., Murdoch, H., Dennis, M.J., Hall, G.A., Bott, R., Crabb, W.D., et al., 2009. Decontamination of prion protein (BSE301V) using a genetically engineered protease. J. Hosp. Infect. 72, 65–70.

Feart, C., Samieri, C., Rondeau, V., Amieva, H., Portet, F., Dartigues, J.F., et al., 2009. Adherence to a Mediterranean diet, cognitive decline and risk of dementia. JAMA 302, 638–648.

Frost, B., Ollesch, J., Wille, H., Diamond, M.I., 2009. Conformational diversity of wild-type tau fibrils specified by template conformation change. J. Biol. Chem. 284, 3546–3551.

Fukushima, T., Tan, X., Luo, Y., Kanda, H., 2010. Relationship between blood levels of heavy metals and Parkinson's disease in China. Neuroepidemiology 34, 18–24.

Gajdusek, D.C., Zigas, V., 1957. Degenerative disease of the central nervous system in New Guinea; the endemic occurrence of kuru in the native population. N. Engl. J. Med. 257, 974–978.

Gasperini, L., Legname, G., 2014. Prion protein and aging. Front. Cell Dev. Biol. 2, 99–102.

Gasperini, L., Meneghetti, E., Pastore, B., Benetti, F., Legname, G., 2015. Prion protein and copper cooperatively protect neurons by modulating NMDA receptor through S-nitrosylation. Antioxid. Redox. 22, 772–784.

Gonzalez, R., Ballester, I., Lopez-Posadas, R., Suarez, M.D., Zarzuelo, A., Martinez-Augustin, O., et al., 2011. Effects of flavonoids and other polyphenols on inflammation. Crit. Rev. Food Sci. Nutr. 51, 331–362.

Gorell, J.M., Johnson, C.C., Rybicki, B.A., Peterson, E.L., Kortsha, G.X., Brown, G.G., et al., 1997. Occupational exposures to metals as risk factors for Parkinson's disease. Neurology 48, 650–658.

Gotsis, E., Anagnostis, P., Mariolis, A., Vlachou, A., Katsiki, N., Karagiannis, A., 2015. Health benefits of the Mediterranean diet: an update of research over the last 5 years. Angiology 66, 304–318.

Gough, K.C., Maddison, B.C., 2010. Prion transmission: prion excretion and occurrence in the environment. Prion 4, 275–282.

Hafner-Bratkovic, I., Gaspersic, J., Smid, L.M., Bresjanac, M., Jerala, R., 2008. Curcumin binds to the alpha-helical intermediate and to the amyloid form of prion protein—a new mechanism for the inhibition of PrPSc accumulation. J. Neurochem. 104, 1553–1564.

Haley, N.J., Seelig, D.M., Zabel, M.D., Telling, G.C., Hoover, E.A., 2009. Detection of CWD prions in urine and saliva of deer by transgenic mouse bioassay. PLoS One 4, e4848.

Haley, N.J., Mathiason, C.K., Carver, S., Zabel, M., Telling, G.C., Hoover, E.A., 2011. Detection of chronic wasting disease prions in salivary, urinary, and intestinal tissues of deer: potential mechanisms of prion shedding and transmission. J. Virol. 85, 6309–6318.

Hansen, C., Angot, E., Bergstrom, A.L., Steiner, J.A., Pieri, L., Paul, G., et al., 2011. alpha-Synuclein propagates from mouse brain to grafted dopaminergic neurons and seeds aggregation in cultured human cells. J. Clin. Invest. 121, 715–725.

Holcomb, L.A., Dhanasekaran, M., Hitt, A.R., Young, K.A., Riggs, M., Manyam, B.V., 2006. Bacopa monniera extract reduçes amyloid levels in PSAPP mice. J. Alzheimers Dis. 9, 243–251.

Hui, Z., Doi, H., Kanouchi, H., Matsuura, Y., Mohri, S., Nonomura, Y., 2004. Alkaline serine protease produced by Streptomyces sp. degrades PrP(Sc). Biochem. Biophys. Res. Commun. 321, 45–50.

Ingolfsson, H.I., Thakur, P., Herold, K.F., Hobart, E.A., Ramsey, N.B., Periole, X., et al., 2014. Phytochemicals perturb membranes and promiscuously alter protein function. ACS Chem. Biol. 9, 1788–1798.

Jackson, G.S., Murray, I., Hosszu, L.L., Gibbs, N., Waltho, J.P., Clarke, A.R., et al., 2001. Location and properties of metal-binding sites on the human prion protein. Proc. Natl. Acad. Sci. U. S. A. 98, 8531–8535.

Jacobs Jr., D.R., Gross, M.D., Tapsell, L.C., 2009. Food synergy: an operational concept for understanding nutrition. Am. J. Clin. Nutr. 89, 1543S–1548S.

Jensen, W.A., McLaren, A.D., 1960. Uptake of proteins by plant cells—the possible occurrence of pinocytosis in plants. Exp. Cell Res. 19, 414–417.

Jeong, J.K., Moon, M.H., Bae, B.C., Lee, Y.J., Seol, J.W., Kang, H.S., et al., 2012. Autophagy induced by resveratrol prevents human prion protein-mediated neurotoxicity. Neurosci. Res. 73, 99–105.

Johnson, C.J., Phillips, K.E., Schramm, P.T., McKenzie, D., Aiken, J.M., Pedersen, J.A., 2006. Prions adhere to soil minerals and remain infectious. PLoS Pathog. 2, e32.

Johnson, C.J., Pedersen, J.A., Chappell, R.J., McKenzie, D., Aiken, J.M., 2007. Oral transmissibility of prion disease is enhanced by binding to soil particles. PLoS Pathog. 3, e93.

Johnson, C.J., Bennett, J.P., Biro, S.M., Duque-Velasquez, J.C., Rodriguez, C.M., Bessen, R.A., et al., 2011. Degradation of the disease-associated prion protein by a serine protease from lichens. PLoS One 6, e19836.

Khan, N., Afaq, F., Saleem, M., Ahmad, N., Mukhtar, H., 2006. Targeting multiple signaling pathways by green tea polyphenol (−)-epigallocatechin-3-gallate. Cancer Res. 66, 2500–2505.

Kim, H.S., Quon, M.J., Kim, J., 2014. New insights into the mechanisms of polyphenols beyond antioxidant properties; lessons from the green tea polyphenol, epigallocatechin 3-gallate. Redox Biol. 2, 187–195.

Kim, H.J., Choi, H.S., Park, J.H., Kim, M.J., Lee, H.G., Petersen, R.B., et al., 2017. Regulation of RhoA activity by the cellular prion protein. Cell Death Dis. 8, e2668.

Landete, J.M., 2012. Updated knowledge about polyphenols: functions, bioavailability, metabolism, and health. Crit. Rev. Food Sci. Nutr. 52, 936–948.

Langeveld, J.P., Wang, J.J., Van de Wiel, D.F., Shih, G.C., Garssen, G.J., et al., 2003. Enzymatic degradation of prion protein in brain stem from infected cattle and sheep. J. Infect. Dis. 188, 1782–1789.

Lauren, J., Gimbel, D.A., Nygaard, H.B., Gilbert, J.W., Strittmatter, S.M., 2009. Cellular prion protein mediates impairment of synaptic plasticity by amyloid-beta oligomers. Nature 457, 1128–1132.

Lawson, V.A., Stewart, J.D., Masters, C.L., 2007. Enzymatic detergent treatment protocol that reduces protease-resistant prion protein load and infectivity from surgical-steel monofilaments contaminated with a human-derived prion strain. J. Gen. Virol. 88, 2905–2914.

Lee, S.T., Chu, K., Jung, K.H., Ban, J.J., Im, W.S., Jo, H., et al., 2015. Altered expression of miR-202 in cerebellum of multiple-system atrophy. Mol. Neurobiol. 51, 180–186.

Limpeanchob, S., Jaipan, S., Rattanakaruna, S., Phrompittayarat, W., Ingkaninan, K., 2008. Neuroprotective effect of Bacopa monnieri on beta-amyloid-induced cell death in primary cortical culture. J. Ethnopharmacol. 120, 112–117.

Lin, C.F., Yu, K.H., Jheng, C.P., Chung, R., Lee, C.I., 2013. Curcumin reduces amyloid fibrillation of prion protein and decreases reactive oxidative stress. Pathogens 2, 506–519.

Maddison, B.C., Rees, H.C., Baker, C.A., Taema, M., Bellworthy, S.J., Thorne, L., et al., 2010. Prions are secreted into the oral cavity in sheep with preclinical scrapie. J. Infect. Dis. 201, 1672–1676.

Malishev, R., Nandi, S., Kolusheva, S., Shaham-Niv, S., Gazit, E., Jelinek, R., 2016. Bacoside-A, an anti-amyloidnatural substance, inhibits membrane disruption by the amyloidogenic determinant of prion protein through accelerating fibril formation. Biochim. Biophys. Acta 1858, 2208–2214.

Manach, C., Scalbert, A., Morand, C., Remesy, C., Jimenez, L., 2004. Polyphenols: food sources and bioavailability. Am. J. Clin. Nutr. 79, 727–747.

Mandel, S.A., Amit, T., Weinreb, O., Youdim, M.B., 2011. Understanding the broad-spectrum neuroprotective action profile of green tea polyphenols in aging and neurodegenerative diseases. J. Alzheimers Dis. 25, 187–208.

Martins, A., Araujo, M.E., Branco, I., Meireles, M., Almeida, J., Sepulveda, C., et al., 2009. Evaluation of plant extracts against prion diseases and cancer. In: Proceedings of the 42nd IUPAC Congress—Chemistry Solutions, Session for Chemistry for Health—Chemistry in the Food Chain and Health, Glasgow, UK, pp. 2–7.

Mathiason, C.K., Powers, J.G., Dahmes, S.J., Osborn, D.A., Miller, K.V., Warren, R.J., et al., 2006. Infectious prions in the saliva and blood of deer with chronic wasting disease. Science 314, 133–136.

Mathiason, C.K., Hays, S.A., Powers, J., Hayes-Klug, J., Langenberg, J., Dahmes, S.J., et al., 2009. Infectious prions in pre-clinical deer and transmission of chronic wasting disease solely by environmental exposure. PLoS One 4, e5916.

McLaren, A.D., Jensen, W.A., Jacobson, L., 1960. Absorption of enzymes and other proteins by barley roots. Plant Physiol. 35, 549–556.

Miller, M.W., Williams, E.S., Hobbs, N.T., Wolfe, L.L., 2004. Environmental sources of prion transmission in mule deer. Emerg. Infect. Dis. 10, 1003–1006.

Mitteregger, G., Vosko, M., Krebs, B., Xiang, W., Kohlmannsperger, V., Nolting, S., et al., 2007. The role of the octa repeat region in neuroprotective function of the cellular prion protein. Brain Pathol. 17, 174–183.

Miura, M., Yoda, M., Takaku, N., Hirose, T., Takeuchi, H., 2007. Clustered negative charges on the lipid membrane surface induce β-sheet formation of prion protein fragment 106–126. Biochemistry 46, 11589–11597.

Mizrahi, M., Friedman-Levi, Y., Larush, L., Frid, K., Binyamin, O., Dori, D., et al., 2014. Pomegranate seed oil nanoemulsions for the prevention and treatment of neurodegenerative diseases: the case of genetic CJD. Nanomedicine 10, 1353–1363.

Nonaka, T., Watanabe, S.T., Iwatsubo, T., Hasegawa, M., 2010. Seeded aggregation and toxicity of {alpha}-synuclein and tau: cellular models of neurodegenerative diseases. J. Biol. Chem. 285, 34885–34898.

Obulesu, M., 2021. Turmeric and Curcumin in Neurodegenerative Diseases, first ed. Elsevier.

Obulesu, M., Dendi,V.S., 2020. Neuroprotective role of resveratrol in Alzheimer's disease. In: Phytomedicine in Alzheimer's Disease, first ed.Taylor and Francis (CRC Press), p. 2020.

Obulesu, M.,Venu, R., Somashekhar, R., 2011. Lipid peroxidation in Alzheimer's Disease: emphasis on metal mediated neurotoxicity. Acta Neurol. Scand. 124, 295–301.

Ono, K., Yoshiike, Y., Takashima, A., Hasegawa, K., Naiki, H., Yamada, M., 2003. Potent anti-amyloidogenic and fibril-destabilizing effects of polyphenols in vitro: implications for the prevention and therapeutics of Alzheimer's disease. J. Neurochem. 87, 172–181.

Pandey, K.B., Rizvi, S.I., 2009. Plant polyphenols as dietary antioxidants in human health and disease. Oxid. Med. Cell. Longev. 2, 270–278.

Paungfoo-Lonhienne, C., Lonhienne, T.G., Rentsch, D., Robinson, N., Christie, M., Webb, R.I., et al., 2008. Plants can use protein as a nitrogen source without assistance from other organisms. Proc. Natl. Acad. Sci. U. S. A. 105, 4524–4529.

Peggion, C., Bertoli, A., Sorgato, M.C., 2016. Almost a century of prion protein(s): from pathology to physiology, and back to pathology. Biochem. Biophys. Res. Commun., 31240–31242. pii:S0006291X.

Pelletier, A., Barul, C., Feart, C., Helmer, C., Bernard, C., Periot, O., et al., 2015. Mediterranean diet and preserved brain structural connectivity in older subjects. Alzheimers Dement. 9, 1023–1031.

Perez-Hernandez, J., Zaldivar-Machorro, V.J., Villanueva-Porras, D., Vega-Avila, E., Chavarria, A., 2016. A potential alternative against neurodegenerative diseases: phytodrugs. Oxid. Med. Cell. Longev. 2016, 8378613. Hindawi Publishing Corporation.

Perron, N.R., Brumaghim, J.L., 2009. A review of the antioxidant mechanisms of polyphenol compounds related to iron binding. Cell Biochem. Biophys. 53, 75–100.

Pignatelli, P., Ghiselli, A., Buchetti, B., 2006. Polyphenols synergistically inhibit oxidative stress in subjects given red and white wine. Atherosclerosis 188, 77–83.

Pilon, J.L., Nash, P.B., Arver, T., Hoglund, D., Vercauteren, K.C., 2009. Feasibility of infectious prion digestion using mild conditions and commercial subtilisin. J. Virol. Methods 161, 168–172.

Porzoor, A., Alford, B., Hügel, H.M., Grando, D., Caine, J., Macreadie, I., 2015. Anti-amyloidogenic properties of some phenolic compounds. Biomol. Ther. 5, 505–527.

Prasad, K.N., Bondy, S.C., 2018. Oxidative and inflammatory events in prion diseases: can they be therapeutic targets? Curr. Aging Sci. 11, 216–225.

Prusiner, S.B., 1982. Novel proteinaceous infectious particles cause scrapie. Science 216, 136–144.

Prusiner, S.B., 2001. Shattuck lecture—neurodegenerative diseases and prions. N. Engl. J. Med. 344, 1516–1526.

Prusiner, S.B., 2012. Cell biology. A unifying role for prions in neurodegenerative diseases. Science 336, 1511–1513.

Rachidi, W., Vilette, D., Guiraud, P., Arlotto, M., Riondel, J., Laude, H., et al., 2003. Expression of prion protein increases cellular copper binding and antioxidant enzyme activities but not copper delivery. J. Biol. Chem. 278, 9064–9072.

Ramassamy, C., 2006. Emerging role of polyphenolic compounds in the treatment of neurodegenerative diseases: a review of their intracellular targets. Eur. J. Pharmacol. 545, 51–64.

Rauter, A.P., Branco, I., Lopes, R.G., Justin, J., Silva, F.V.M., Noronha, J.P., 2007. A new lupene triterpenetriol and anticholinesterase activity of Salvia sclareoides. Fitoterapia 78, 474–481.

Rauter, A.P., Dias, C., Martins, A., Branco, I., Neng, N.R., Nogueira, J.M., et al., 2012. Non-toxic Salvia sclareoides Brot. extracts as a source of functional food ingredients: phenolic profile, antioxidant activity and prion binding properties. Food Chem. 132, 1930–1935.

Rice-Evans, C.A., Miller, N.J., Paganga, G., 1996. Structure antioxidant activity relationships of flavonoids and phenoliacids. Free Radic. Biol. Med. 20, 933–956.

Rybicki, B.A., Johnson, C.C., Uman, J., Gorell, J.M., 1993. Parkinson's disease mortality and the industrial use of heavy metals in Michigan. Mov. Disord. 8, 87–92.

Sanchez, M.T.B., de Matos, A.M., Rauter, A.P., 2017. Exploring anti-prion glyco-based and aromatic scaffolds: a chemical strategy for the quality of life. Molecules 22, 864.

Scarmeas, N., Stern, Y., Tang, M.X., Mayeux, R., Luchsinger, J.A., 2006. Mediterranean diet and risk for Alzheimer's disease. Ann. Neurol. 59, 912–921.

Scarmeas, N., Stern, Y., Mayeux, R., Manly, J.J., Schupf, N., Luchsinger, J.A., 2009. Mediterranean diet and mild cognitive impairment. Arch. Neurol. 66, 216–225.

Schneider, C., Gordon, O.N., Edwards, R.L., Luis, P.B.J., 2015. Degradation of curcumin: from mechanism to biological implications. Agric. Food Chem. 63, 7606–7614.

Seidel, B., Thomzig, A., Buschmann, A., Groschup, M.H., Peters, R., Beekes, M., et al., 2007. Scrapie agent (strain 263K) can transmit disease via the oral route after persistence in soil over years. PLoS One 2, e435.

Sergent, T., Piront, N., Meurice, J., Toussaint, O., Schneider, Y.J., 2010. Anti-inflammatory effects of dietary phenolic compounds in an in vitro model of inflamed human intestinal epithelium. Chem. Biol. Interact. 188, 659–667.

Shi, F., Yang, Y., Wang, T., Kouadir, M., Zhao, D., Hu, S., 2016. Cellular prion protein promotes neuronal differentiation of adipose-derived stem cells by upregulating miRNA-124. J. Mol. Neurosci. 59, 48–55.

Singh, M., Arseneault, M., Sanderson, T., Murthy, V., Ramassamy, C., 2008. Challenges for research on polyphenols from foods in Alzheimer's disease: bioavailability, metabolism, cellular and molecular mechanisms. J. Agric. Food Chem. 56, 4855–4873.

Singh, B., Parsaik, A.K., Mielke, M.M., Erwin, P.J., Knopman, D.S., Petersen, R.C., et al., 2014. Association of Mediterranean diet with mild cognitive impairment and Alzheimer's disease: a systematic review and meta-analysis. J. Alzheimers Dis. 39, 271–282.

Slapsak, U., Salzano, G., Amin, L., Abskharon, R.N., Ilc, G., Zupancic, B., et al., 2016. The N terminus of the prion protein mediates functional interactions with the neuronal cell adhesion molecule (NCAM) fibronectin domain. Biol. Chem. 291, 21857–21868.

Suganuma, M., Okabe, S., Oniyama, M., Tada, Y., Ito, H., Fujiki, H., 1998. Wide distribution of [3H](−)-epigallocatechin gallate, a cancer preventive tea polyphenol, in mouse tissue. Carcinogenesis 19, 1771–1776.

Tachibana, H., Koga, K., Fujimura, Y., Yamada, K., 2004. A receptor for green tea polyphenol EGCG. Nat. Struct. Mol. Biol. 11, 380–381.

Tagliavini, F., Prelli, F., Verga, L., Giaccone, G., Sarma, R., Gorevic, P., et al., 1993. Synthetic peptides homologous to prion protein residues 106–147 form amyloid-like fibrils *in vitro*. Proc. Natl. Acad. Sci. U. S. A. 90, 9678–9682.

Tamguney, G., Miller, M.W., Wolfe, L.L., Sirochman, T.M., Glidden, D.V., Palmer, C., et al., 2009. A symptomatic deer excrete infectious prions in faeces. Nature 461, 529–532.

Terry, L.A., Howells, L., Bishop, K., Baker, C.A., Everest, S., Thorne, L., et al., 2011. Detection of prions in the faeces of sheep naturally infected with classical scrapie. Vet. Res. 42, 65.

Toni, M., Massimino, M.L., De Mario, A., Angiulli, E., Spisni, E., 2017. Metal dyshomeostasis and their pathological role in prion and prion like diseases: the basis for a nutritional approach. Front. Neurosci. 11, 3.

Tsao, R., 2010. Chemistry and biochemistry of dietary polyphenols. Nutrients 2, 1231–1246.

Viviane, L., Ngoungoure, N., Schluesener, J., Moundipa, P.F., Schluesener, H., 2015. Natural polyphenols binding to amyloid: a broad class of compounds to treat different human amyloid diseases. Mol. Nutr. Food Res. 59, 8–20.

Walle, T., 2011. Bioavailability of resveratrol. Ann. N.Y. Acad. Sci. 1215, 9–15.

Walter, E.D., Stevens, D.J., Visconte, M.P., Millhauser, G.L., 2007. The prion protein is a combined zinc and copper binding protein: Zn^{2+} alters the distribution of Cu^{2+} coordination modes. J. Am. Chem. Soc. 129, 15440–15441.

Wang, J., Zhang, J., Shi, Q., Zhang, B.Y., Chen, C., Chen, L.N., et al., 2015. Scrapie infection in experimental rodents and SMBS15 cells decreased the brain endogenous levels and activities of Sirt1. J. Mol. Neurosci. 55, 1022–1030.

Wang, J., Zhang, B.Y., Zhang, J., Xiao, K., Chen, L.N., Wang, H., et al., 2016. Treatment of SMB-S15 cells with resveratrol efficiently removes the PrPSc accumulation *in vitro* and prion infectivity *in vivo*. Mol. Neurobiol. 53, 5367–5376.

White, A.R., Collins, S.J., Maher, F., Jobling, M.F., Stewart, L.R., Thyer, J.M., et al., 1999. Prion protein deficient neurons reveal lower glutathione reductase activity and increased susceptibility to hydrogen peroxide toxicity. Am. J. Pathol. 155, 1723–1730.

Witte, A.V., Kerti, L., Margulies, D.S., Floel, A., 2014. Effects of resveratrol on memory performance, hippocampal functional connectivity and glucose metabolism in healthy older adults. J. Neurosci. 34, 7862–7870.

Yamamoto, N., Matsuzaki, K., Yanagisawa, K., 2005. Cross-seeding of wild-type and hereditary variant-type amyloid beta-proteins in the presence of gangliosides. J. Neurochem. 95, 1167–1176.

Yang, F., Lim, G.P., Begum, A.N., Ubeda, O.J., Simmons, M.R., Ambegaokar, S.S., et al., 2005. Curcumin inhibits formation of amyloid beta oligomers and fibrils, binds plaques, and reduces amyloid in vivo. J. Biol. Chem. 280, 5892–5901.

Yoshioka, M., Murayama, Y., Miwa, T., Miura, K., Takata, M., Yokoyama, T., et al., 2007. Assessment of prion inactivation by combined use of bacillus-derived protease and SDS. Biosci. Biotechnol. Biochem. 71, 2565–2568.

Zayed, J., Ducic, S., Campanella, G., Panisset, J.C., Andre, P., Masson, H., et al., 1990. Environmental factors in the etiology of Parkinson's disease. Can. J. Neurol. Sci. 17, 286–291.

Zhang, Q., Tu, T., D'Avignon, D.A., Gross, M.L., 2009. Balance of beneficial and deleterious health effects of quinones: a case study of the chemical properties of genistein and estrone quinones. J. Am. Chem. Soc. 131, 1067–1076.

CHAPTER 4

Efficacy of plant extracts against Friedreich's ataxia

Introduction

Pathology

German neurologist Nicholaus Friedreich, in a family of professors of medicine, described a new disease in consecutive papers published between 1863 and 1877 (Koeppen, 2013). Therefore, the new disease was named Friedreich's ataxia (FA). The disease is prevalent in individuals of Indo-European and Afro-Asiatic origin, while the frequency on Amerindians, sub-Saharan, Chinese, Japanese, and Southeast Asiatic populations is minimal (Labuda et al., 2000; Rodriguez et al., 2020). FA is the most frequent subtype of autosomal recessive cerebellar ataxias after ataxia-telangiectasia (AT) and ataxia oculomotor apraxia type 1 (AOA1) and type 2 (AOA2) (Bird, 2019; Phang et al., 2021). The most frequent X-linked ataxia is weak X-linked tremor ataxia syndrome (FXTAS). The age of onset differs depending on the type of hereditary ataxia (HA), but onset facilitates the diagnosis of HA. Infancy onset indicates AT; childhood or teenage onset indicates EA, FA, AOA1, and AOA2; and adult onset indicates SCA, FXTAS, or other autosomal recessive cerebellar ataxias (Beaudin et al., 2019; Phang et al., 2021).

FA is an autosomal recessive disorder induced by the abnormal expansion of GAA repeats resulting in inhibited expression of FXN protein (Jauslin et al., 2007). This disorder upsets central and peripheral nervous systems, the heart, and endocrine pancreas (Llorens et al., 2019). Expansion of GAA repeats or point mutations in the gene FXN located on chromosome 9 has been observed in FA (Chamberlain et al., 1988). Several mutations were found on the first intron of FXN with 8–22 expansions taking place in normal alleles. While a few patients have an expansion of GAA repeats on one allele and point mutations on the other allele, the majority have repeat expansions on both alleles. Patients having substantial expansions experience early onset, and more complications and disease symptoms. There is a radical decrease in the levels of FXN in patients (Campuzano et al., 1996).

Plant Extracts in Neurodegenerative Diseases
https://doi.org/10.1016/B978-0-323-95762-5.00006-0

The onset of the disease is generally observed during adolescence, and symptoms such as unsteady gait, lack of lower limb reflexes, dysarthria, dysphagia, eye movement defects, scoliosis, foot malformations, cardiomyopathy, and the presence of pyramidal signs (neural pathways controlling voluntary motor functions) have been observed (Sbodio et al., 2019).

Frataxin

FXN, a low-molecular-weight protein of 23 kDa, is found in mitochondria (Koutnikova et al., 1997). In yeast, deletion of the FXN homologue, YFH1, promotes iron aggregation, deprivation of mtDNA, and sensitivity to H_2O_2 and iron (Babcock et al., 1997; Foury and Cazzalini, 1997; Wilson and Roof, 1997). In mice, knockout of the FXN gene induces embryonic mortality, but a few studies showed no change in iron aggregation, whereas a few have noticed increased iron levels (Cossee et al., 2000, Puccio et al., 2001). Reduction in the iron/sulfur (Fe-S) cluster-containing subunits of mitochondrial respiratory complexes I, II, and III has been shown in endomyocardial biopsies of two FA patients, suggesting a role of FXN in Fe-S protein homeostasis (Rotig et al., 1997). Additionally, activity of the Fe-S protein aconitase, which plays a role in iron homeostasis, is decreased in FA patients. Impaired iron metabolism and mitochondrial respiration eventually lead to increased oxidative stress due to iron accumulation and its role in Fenton reaction to produce reactuve oxygen species (ROS). There has been an ambiguity in the role of iron in FA pathogenesis since it is an inconsistent characteristic in FA cell culture and mouse models, and has been a late episode when Fe-S proteins are minimal (Bayot et al., 2011; Cossee et al., 2000). Nevertheless, *Drosophila* and CRISPR mouse models devoid of FXN have been found to aggregate iron and induce toxicity (Chen et al., 2016a,b). Knockout of mouse FXN in the central nervous system (CNS) with AAV and CRISPR/Cas9 induces behavioral and neurological phenotypes akin to the previous reports showing neuronal conditional knockout mouse models such as smaller body size, hunchback phenotype, perturbed locomotion, and reduced lifespan (Puccio et al., 2001). Intra- and extracellular iron aggregation (both Fe^{2+} and Fe^{3+}) have been noticed. Iron aggregation has been observed to initiate sphingolipid synthesis and activation of 3-phosphoinositide-dependent protein kinase-1 and myocyte enhancer factor-2 (Mef2). Mef2 then triggers transcription of downstream target genes that regulate degeneration (Chen et al., 2016b). A few other studies have shown that FXN diminution results in increased sensitivity to oxidative stress. Fibroblasts obtained from FA patients demonstrated a dulled response to enhanced iron levels. The $O_2^{\bullet-}$-detoxifying mitochondrial enzyme MnSOD could not react in FA fibroblasts exposed to iron (Jiralerspong et al., 2001). This defect, which shows reduction of PGC1α, a protein responsible for mitochondrial energetics and a coactivator for PPARγ in FA cells, results in suboptimal functioning of antioxidant defense pathways (Marmolino et al., 2010). Based on these findings, the hypersensitivity of fibroblasts in FA patients to oxidative stress can be averted by decreasing lipid peroxidation with deuterated poly unsaturated fatty acids (PUFAs) or by initiating the Nrf2 pathway (Abeti et al., 2015, 2016). Therefore, it can be understood that increased oxidative stress results in the pathology of a range of ataxias, which significantly impair cerebellar function. Antioxidant therapy in these ataxias exhibited different efficacy and warranted further studies (Barca et al., 2019).

FXN deficit in FA hampers the intracellular antioxidant defense system leading to reduced glutathione (GSH) levels (Piemonte et al., 2001; Auchere et al., 2008; Clark et al., 2018) and expressions of antioxidant enzymes such as SOD, CAT, glutaredoxin, and thioredoxin (Gomes and Santos, 2013). In addition, a high amount of ROS is produced by an iron-dependent Fenton reaction and mitochondrial impairment. The impaired antioxidant defense system and ROS aggregation result in uncontrolled enhancement in the oxidative damage, eventually contributing to DNA damage and cell death (Santos et al., 2010; Lefevre et al., 2012; Carletti and Piemonte, 2014).

Iron accumulation

Iron accumulation is a primary pathological event in FA, and it has been found to be provoked by FXN deficit. The first iron accumulation and its association with FA were carried out in postmortem heart samples of a few patients that showed intense intracytoplasmic aggregation of iron particles in the cardiac fibers (Sanchez-Casis et al., 1976; Lamarche et al., 1980). These results were further confirmed in succeeding studies that employed techniques such as Perls Prussian blue staining for histological identification or atomic absorption spectroscopy to calculate total iron levels in tissue. The increased iron in the heart of FA patients found as multifocal aggregates was intensely contained instead of a diffuse pattern (Ramirez et al., 2012; Koeppen et al., 2015). This regional allocation of increased iron could be accountable for the findings of iron levels comparable to healthy controls in bulk extracts or from casual biopsied cardiac samples (Michael et al., 2006; Koeppen et al., 2015; Kruger et al., 2016). Additionally, positive iron staining was noticed in the liver and spleen of patients, demonstrating a distribution constant with a mitochondrial location (Bradley et al., 2000) as has been explained in the FXN-deficient yeast model (Babcock et al., 1997; Foury and Cazzalini, 1997).

Alterations in the distribution of iron and other metals like copper and zinc in the dentate nuclei of FA patients have also been found (Koeppen et al., 2012), although the total quantity of these metals was equal to the controls. It can also be emphasized that copper can also produce oxygen radicals and in FA may worsen the oxidative damage induced by the enhanced quantity of iron. Mitochondrial iron aggregation has also been reported in fibroblasts and lymphoblasts of FA patients. Despite the discovery of enhanced total iron content (Delatycki et al., 1999) or labile iron fraction (Tan et al., 2001), normal iron levels were noticed in other studies (Wong et al., 1999; Sturm et al., 2005).

Iron dyshomeostasis in Friedreich's ataxia

Despite the unclear understanding of the mechanisms of mitochondrial iron loading in FA, a wealth of studies has explained the hypotheses that could describe this modification. Huang et al. (2009) observed the changed heme and iron-sulfur cluster synthesis and mitochondrial iron storage in the MCK-KO mouse model. Profoundly decreased activity and expression of ISC enzymes, like NFS1 and ISCU1/2, were noticed that were in accordance with the decrease in the expression of succinate dehydrogenase complex, subunit A (Sdha) (Huang et al., 2009). Simultaneously, other studies have reported a downregulation in a few heme synthesis enzymes and reduced heme levels in the heart of the MCK KO model (Schoenfeld et al., 2005; Huang et al., 2009).

A reduction in the synthesis of various products that need iron in the mitochondria could describe the aggregation of this metal since it cannot be transported from the organelle. Nevertheless, frataxin deficit can also influence the expression of a plethora of proteins, a few of them vital for the regulation of iron metabolism. In dentate nuclei, Koeppen et al. (2007) noticed alteration of the expression of proteins like transferrin receptor 1 (TFR1), ferritins (FRTs), and ferroportin (FPN), which were modifications distributed among animal models of FA. Martelli and Puccio (2014) reported significant TFR1 upregulation in the MCK-KO mouse, resulting in enhanced iron uptake and downregulation of FPN, reducing cellular iron release from the cytosol.

Plant extracts

Studies have shown the antioxidant activities of the aqueous extract of Hericiumerinaceusin, an in vitro model of FA comprising L-buthionine sulfoximine (L-BSO)-initiated human dermal fibroblast expressing unusual expansion of GAA triplet repeat (Lew et al., 2020). L-BSO is an inhibitor of γ-glutamyl cysteine synthetase, which is an essential component in GSH biogenesis. Cells treated with L-BSO have decreased levels of GSH, which enhanced the vulnerability of cells to oxidative damage (Takahashi et al., 2010). Treatment of cells with the extract (32 mg/mL) for 24 h reinstated the GSH reduction, as evidenced by an enhanced GSH/GSSG ratio. The extract also curtailed the lactate dehydrogenase (LDH) level and apoptotic body formation, as observed by LDH assay and Hoechst 33258 staining, respectively. The findings recommended that the extract was capable of avoiding oxidative damage-initiated cell death by averting GSH depletion (Lew et al., 2020).

Rutin, resveratrol, and vitamin E that counter ROS-mediated cascades like JNK and NF-kB showed protection against neurodegeneration both in vitro and in vivo (Zuo et al., 2015). Vitamin E substantially protected against oxidative stress than vitamins A and C in a rat brain (Zaidi and Banu, 2004). In line with this, a 2-year introduction of vitamin E at a dose of 2000 IU per day has been found to decrease the functional deficit linked to AD (Sano et al., 1997). The combination of vitamin E and coenzyme Q10 ameliorates energy production in a few cases of FA by inhibiting oxidative stress and reinstating mitochondrial function (Lodi et al., 2001; Liu et al., 2018).

Coenzyme Q10 and idebenone

Coenzyme Q10 or ubiquinone, a small lipophilic molecule present in the inner mitochondrial membrane, exhibits antioxidant activity and averts the oxidation of cellular structures and regulates other antioxidant molecules like vitamins C and E (Bentinger et al., 2010; Orsucci et al., 2011). Ubiquinone or its derivates exerted protective efficacy in fibroblasts from FA patients (Jauslin et al., 2003). Nearly 20 years ago, an open-label study enrolling 10 FA patients was carried out to evaluate the impact of coenzyme

Q10-vitamin E as an antioxidant-combined therapy. The conclusion parameters were ICARS evaluation and echocardiography. After 6 months, participants failed to show constant clinical benefits (Lodi et al., 2001). The authors noticed signs of bioenergetic enhancement and ascribed the short study duration as the reason for the absence of effects, so they continued to monitor the same patients.

This long-term longitudinal follow-up study examined whether the perfection was continued for 47 months with the same treatment. There was remarkable amelioration in cardiac and skeletal muscle bioenergetics when compared to cross-sectional data from 77 FA subjects. Despite the continued deficit in posture and gait symptoms, the ICARS score showed positive results (Hart et al., 2005). After a few years, the same authors conducted a double-blind randomized trial in 50 FA patients for 2 years. Participants were categorized into high- and low-dose groups of coenzyme Q10-vitamin E combined therapy. Surprisingly, 49% of the patients who were monitored until the end of the study (21 of 43) exhibited deliberate degeneration and clinical amelioration resulting in an enhanced ICARS score. This data has accentuated a robust correlation between efficacy of the therapy and reduced baseline coenzyme Q10 and vitamin E levels (Cooper et al., 2008).

Nuclear factor erythroid 2-related factor

Nuclear factor erythroid 2-related factor (NRF2) signaling impairment has been extensively studied in animal models of FA. An FXN knockout mouse line exhibited reduced NRF2 expression and enhanced KEAP1 expression (Anzovino et al., 2017); in the mouse motor neuron cell line NSC-34, FXN shRNA similarly decreased NRF2 expression and activity (Paupe et al., 2009). Conversely, initiating NRF2 with sulforaphane or the NRF2-activating compound EPI-7443 in neural stem cells extracted from an FA mouse model was found to restore correct differentiation, which was earlier futile in those cells (La Rosa et al., 2019). Sulforaphane also enhanced FXN levels and neurite number and extension in cultured motor neurons (Petrillo et al., 2017). Similarly, a decrease in NRF2 expression has been found in fibroblasts isolated from patients (Paupe et al., 2009; Petrillo et al., 2017), and sulforaphane treatment enhanced FXN levels and increased neurite outgrowth in these cells (Petrillo et al., 2017).

RTA-408, which is also called omavelaxolone, is synthetically derived from oleanolic acid, a natural triterpenoid abundantly available in food and plants (Probst et al., 2015; Indelicato and Bosch, 2018). An NRF2-initiating compound, omavelaxolone has been found to be advantageous in the KIKO and the YG8R mouse models of FA and in fibroblasts isolated from human patients. Omavelaxolone reinstated complex I activity, enhanced glutathione levels, reduced ROS, and reestablished the mitochondrial membrane potential averting cell death in these models (Abeti et al., 2018). Depending on these substantial preclinical studies, omaveloxolone is presently under clinical testing.

It has been reported that the results of a phase II trial showed appropriate tolerance to the compound after 48 weeks of treatment and led to substantially enhanced scores on the modified FA rating scale (Reata Pharmaceuticals, 2019).

Resveratrol

Resveratrol is a natural polyphenol with potential antioxidant, anticancer, and neuroprotective efficacy (Vang et al., 2011). It has been found to upregulate FXN in FA patient-derived cells and consequently in an FA mouse model (Li et al., 2013). Resveratrol-mediated FXN upregulation mechanisms are yet to be understood. A wealth of studies has shown the signaling cascade promoted by resveratrol initiates PGC-1α and Nrf2 pathways (Ren et al., 2011; Yiu et al., 2015). These results, along with an appropriate safety profile (Vang et al., 2011), initiated an open-label trial of resveratrol in FA (Yiu et al., 2015). Twenty-four patients were treated with a low or high dose of the drug (1 or 5 g daily) for 12 weeks; the patients treated with a high dose only showed amelioration of neurological scale (FARS). However, the key end point, i.e., the upregulation of FXN in peripheral blood mononuclear cells (PBMCs), was not accomplished. Besides the short study duration, further drawbacks may have limited the oral bioavailability (Cottart et al., 2014; Yiu et al., 2015).

Antioxidants

A-tocopherylquinone is a semisynthetic compound structurally akin to coenzyme Q10 that has been initially designed to combat inherited mitochondrial disorders (Hawi et al., 2012). It was examined in a 4-week, double-blind, randomized, placebo-control trial carried out in 31 FA patients (Lynch et al., 2012). The important end point was the change in Disposition Index, a biochemical symbol of pancreatic β-cells reparation for insulin resistance. Drug tolerance was found adequate. The primary end point was not accomplished by the end of the study, although a remarkable amelioration in neurological scale (FARS) versus placebo was found. Further studies have not been carried out to discover the efficacy of this compound in the treatment of FA.

Epicatechins are flavonoids obtained from natural cacao. (−)-Epicatechin ameliorated mitochondrial function and biogenesis in various models of cardiac and metabolic diseases (Moreno-Ulloa et al., 2015; Ramirez-Sanchez et al., 2016). (+)-Epicatechin is a substantial synthetic variant of the naturally available (−)-Epicatechin, and it is presently examined as a substantial treatment in muscular dystrophy and FA.

Linoleic acid and analogous, long-chain, polyunsaturated fatty acids (PFAs) are vital components of lipid membranes that are physiologically vulnerable to oxidation by ROS. The substitution of hydrogen (1H) atoms with the isotope deuterium (2H) increases resistance to peroxidation (Cotticelli et al., 2013); therefore, altered deuterated PFAs have been used as neuroprotective compounds. RT001, being a deuterated synthetic homologue of linoleic acid ethyl ester, has been examined against linoleic acid ethyl ester in one randomized, double-blind, controlled trial in 18 FA patients for 4 weeks (a first in humans). Major targets were safety, tolerability, and pharmacokinetics. Secondary requirements were the FARS, a timed foot-walk test, and cardiopulmonary exercise testing. Authors reported that the study accomplished its primary safety and tolerability end points. Promising results were recently reported in a research meeting (Heerinck et al., 2017). An amelioration in peak workload in the RT001 group compared to placebo ($P = .02$) and improvement in the neurological scales in the drug group were noticed; therefore, further studies are planned.

Conclusions and future perspectives

FA is a rare ND with complex etiology. Since FXN plays a vital role in the pathology of FA, researchers are currently focusing on it. Iron dyshomeostasis also has been found to play a key role in the pathology. Therefore, reducing iron overload can also be a viable therapeutic option in the treatment of FA.

Detection of feasibly identifiable substitute end-point measures and disease biomarkers that appropriately are on par with disease history plays a pivotal role in formulating substantial clinical trials in which patients will be grouped in an accurate manner. In line with this, little progress has been accomplished with surrogate tissues like lymphoblasts to measure expression levels of given mRNA or miRNA as substantial markers (Seco-Cervera et al., 2017, 2018; Hayashi et al., 2020) and platelets to monitor appropriate frataxin levels or glucose/lipid metabolic disorders in controls and patients (Worth et al., 2015; Guo et al., 2018; Rodriguez et al., 2020). As discussed in this chapter, a plethora of plant-based antioxidants substantially improve the symptoms of FA. However, a limitation of the current therapeutics is short systemic circulation. Therefore, garnering expertise from several fields such as medicine, biochemistry, and nanotechnology probably places a cornerstone on which robust therapeutic arsenal can be streamlined.

References

Abeti, R., Uzun, E., Renganathan, I., Honda, T., Pook, M.A., Giunti, P., 2015. Targeting lipid peroxidation and mitochondrial imbalance in Friedreich's ataxia. Pharmacol. Res. 99, 344–350.

Abeti, R., Parkinson, M.H., Hargreaves, I.P., Angelova, P.R., Sandi, C., Pook, M.A., et al., 2016. Mitochondrial energy imbalance and lipid peroxidation cause cell death in Friedreich's ataxia. Cell Death Dis. 7, e2237.

Abeti, R., Baccaro, A., Esteras, N., Giunti, P., 2018. Novel Nrf2-inducer prevents mitochondrial defects and oxidative stress in Friedreich's ataxia models. Front. Cell. Neurosci. 12, 188.

Anzovino, A., Chiang, S., Brown, B.E., Hawkins, C.L., Richardson, D.R., Huang, M.L., 2017. Molecular alterations in a mouse cardiac model of Friedreich ataxia: an impaired Nrf2 response mediated via upregulation of Keap1 and activation of the Gsk3beta axis. Am. J. Pathol. 187, 2858–2875.

Auchere, F., Santos, R., Planamente, S., Lesuisse, E., Camadro, J.M., 2008. Glutathione-dependent redox status of frataxin-deficient cells in a yeast model of Friedreich's ataxia. Hum. Mol. Genet. 17, 2790–2802.

Babcock, M., de Silva, D., Oaks, R., Davis-Kaplan, S., Jiralerspong, S., Montermini, L., et al., 1997. Regulation of mitochondrial iron accumulation by Yfh1p, a putative homolog of frataxin. Science, 1709–1712.

Barca, E., Emmanuele, V., DiMauro, S., Toscano, A., Quinzii, C.M., 2019. Anti-oxidant drugs: novelties and clinical implications in cerebellar ataxias. Curr. Neuropharmacol. 17, 21–32.

Bayot, A., Santos, R., Camadro, J.M., Rustin, P., 2011. Friedreich's ataxia: the vicious circle hypothesis revisited. BMC Med. 9, 112.

Beaudin, M., Matilla-Duenas, A., Soong, B.W., Pedroso, J.L., Barsottini, O.G., Mitoma, H., et al., 2019. The classification of autosomal recessive cerebellar ataxias: a consensus statement from the society for research on the cerebellum and ataxias task force. Cerebellum 18, 1098–1125.

Bentinger, M., Tekle, M., Dallner, G., 2010. Coenzyme Q—biosynthesis and functions. Biochem. Biophys. Res. Commun. 396, 74–79.

Bird, T.D., 2019. Hereditary ataxia overview. In: Adam, M.P., Ardinger, H.H., Pagon, R.A., Wallace, S.E. (Eds.), GeneReviews® [Internet]. University of Washington, Seattle.

Bradley, J.L., Blake, J.C., Chamberlain, S., Thomas, P.K., Cooper, J.M., Schapira, A.H., 2000. Clinical, biochemical and molecular genetic correlations in Friedreich's ataxia. Hum. Mol. Genet. 9, 275–282.

Campuzano, V., Montermini, L., Molto, M.D., Pianese, L., Cossee, M., Cavalcanti, F., et al., 1996. Friedreich's ataxia: autosomal recessive disease caused by an intronic GAA triplet repeat expansion. Science 271, 1423–1427.

Carletti, B., Piemonte, F., 2014. Friedreich's ataxia: a neuronal point of view on the oxidative stress hypothesis. Antioxidants (Basel) 3, 592–603.

Chamberlain, S., Shaw, J., Rowland, A., Wallis, J., South, S., Nakamura, Y., et al., 1988. Mapping of mutation causing Friedreich's ataxia to human chromosome 9. Nature 334, 248–250.

Chen, K., Ho, T.S., Lin, G., Tan, K.L., Rasband, M.N., Bellen, H.J., 2016a. Loss of frataxin activates the iron/sphingolipid/PDK1/Mef2 pathway in mammals. Elife 5, e20732.

Chen, K., Lin, G., Haelterman, N.A., Ho, T.S., Li, T., Li, Z., et al., 2016b. Loss of frataxin induces iron toxicity, sphingolipid synthesis, and Pdk1/Mef2 activation, leading to neurodegeneration. Elife 5. pii: e16043.

Clark, E., Johnson, J., Dong, Y.N., Mercado-Ayon, E., Warren, N., Zhai, M., et al., 2018. Role of frataxin protein deficiency and metabolic dysfunction in Friedreich ataxia, an autosomal recessive mitochondrial disease. Neuronal Signals 2. NS20180060.

Cooper, J.M., Korlipara, L.V.P., Hart, P.E., Bradley, J.L., Schapira, A.H.V., 2008. Coenzyme Q10 and vitamin E deficiency in Friedreich's ataxia: predictor of efficacy of vitamin e and coenzyme Q10 therapy. Eur. J. Neurol. 15, 1371–1379.

Cossee, M., Puccio, H., Gansmuller, A., Koutnikova, H., Dierich, A., LeMeur, M., et al., 2000. Inactivation of the Friedreich ataxia mouse gene leads to early embryonic lethality without iron accumulation. Hum. Mol. Genet. 9, 1219–1226.

Cottart, C.H., Nivet-Antoine, V., Beaudeux, J.L., 2014. Review of recent data on the metabolism, biological effects, and toxicity of resveratrol in humans. Mol. Nutr. Food Res. 58, 7–21.

Cotticelli, M.G., Crabbe, A.M., Wilson, R.B., Shchepinov, M.S., 2013. Insights into the role of oxidative stress in the pathology of Friedreich ataxia using peroxidation resistant polyunsaturated fatty acids. Redox Biol. 1, 398–404.

Delatycki, M.B., Camakaris, J., Brooks, H., Evans-Whipp, T., Thorburn, D.R., Williamson, R., et al., 1999. Direct evidence that mitochondrial iron accumulation occurs in Friedreich ataxia. Ann. Neurol. 45, 673–675.

Foury, F., Cazzalini, O., 1997. Deletion of the yeast homologue of the human gene associated with Friedreich's ataxia elicits iron accumulation in mitochondria. FEBS Lett. 411, 373–377.

Gomes, C.M., Santos, R., 2013. Neurodegeneration in Friedreich's ataxia: from defective frataxin to oxidative stress. Oxid. Med. Cell. Longev. 2013, 487534.

Guo, L., Wang, Q., Weng, L., Hauser, L.A., Strawser, C.J., Rocha, A.G., et al., 2018. Liquid chromatography-high resolution mass spectrometry analysis of platelet frataxin as a protein biomarker for the rare disease Friedreich's ataxia. Anal. Chem. 90, 2216–2223.

Hart, P.E., Lodi, R., Rajagopalan, B., Bradley, J.L., Crilley, J.G., Turner, C., et al., 2005. Antioxidant treatment of patients with Friedreich ataxia: four-year follow-up. Arch. Neurol. 62, 621–626.

Hawi, A., Heald, S., Sciascia, T., 2012. Use of an adaptive study design in single ascending-dose pharmacokinetics of A0001 (alpha-tocopherylquinone) in healthy male subjects. J. Clin. Pharmacol. 52, 65–77.

Hayashi, G., Jasoliya, M., Sahdeo, S., Saccà, F., Pane, C., Filla, A., et al., 2020. Dimethyl Fumarate Mediates Nrf2-Dependent Mitochondrial Biogenesis in Mice and Humans. Available from: https://www.ncbi.nlm.nih.gov/pmc/articles/PMC6251607/. (Accessed 23 March 2020).

Heerinck, F., Shaw, J., Shchepinov, M., et al., 2017. RT001 first-in human clinical trial demonstrates safety, favorable pharmacokinetics, and early signals of efficacy in Friedreich's ataxia. In: International Ataxia Research Conference 2017, Pisa, Italy.

Huang, M.L.H., Becker, E.M., Whitnall, M., Suryo Rahmanto, Y., Ponka, P., Richardson, D.R., 2009. Elucidation of the mechanism of mitochondrial iron loading in Friedreich's ataxia by analysis of a mouse mutant. Proc. Natl. Acad. Sci. U. S. A. 106, 16381–16386.
Indelicato, E., Bosch, S., 2018. Emerging therapeutics for the treatment of Friedreich's ataxia. Expert Opin. Orphan Drugs 6 (1), 57–67.
Jauslin, M.L., Meier, T., Smith, R.A.J., Murphy, P.M., 2003. Mitochondria-targeted antioxidants protect Friedreich Ataxia fibroblasts from endogenous oxidative stress more effectively than untargeted antioxidants. FASEB J. 17, 1–10.
Jauslin, M.L., Vertuani, S., Durini, E., Buzzoni, L., Ciliberti, N., Verdecchia, S., et al., 2007. Protective effects of Fe-Aox29, a novel antioxidant derived from a molecular combination of idebenone and vitamin E, in immortalized fibroblasts and fibroblasts from patients with Friedreich ataxia. Mol. Cell. Biochem. 302, 79–85.
Jiralerspong, S., Ge, B., Hudson, T.J., Pandolfo, M., 2001. Manganese superoxide dismutase induction by iron is impaired in Friedreich ataxia cells. FEBS Lett. 509, 101–105.
Koeppen, A.H., 2013. Nikolaus Friedreich and degenerative atrophy of the dorsal columns of the spinal cord. J. Neurochem. 126, 4–10.
Koeppen, A.H., Michael, S.C., Knutson, M.D., Haile, D.J., Qian, J., Levi, S., et al., 2007. The dentate nucleus in Friedreich's ataxia: the role of iron responsive proteins. Acta Neuropathol. 114, 163–173.
Koeppen, A.H., Ramirez, R.L., Yu, D., Collins, S.E., Qian, J., Parsons, P.J., et al., 2012. Friedreich's ataxia causes redistribution of iron, copper, and zinc in the dentate nucleus. Cerebellum 11, 845–860.
Koeppen, A.H., Ramirez, R.L., Becker, A.B., Bjork, S.T., Levi, S., Santambrogio, P., et al., 2015. The pathogenesis of cardiomyopathy in Friedreich ataxia. PLoS One 10, e0116396.
Koutnikova, H., Campuzano, V., Foury, F., Dolle, P., Cazzalini, O., Koenig, M., 1997. Studies of human, mouse and yeast homologues indicate a mitochondrial function for frataxin. Nat. Genet. 16, 345–351.
Kruger, P.C., Yang, K.X., Parsons, P.J., Becker, A.B., Feustel, P.J., Koeppen, A.H., 2016. Abundance and significance of iron, zinc, copper, and calcium in the hearts of patients with Friedreich ataxia. Am. J. Cardiol. 118, 127–131.
La Rosa, P., Russo, M., D'Amico, J., Petrillo, S., Aquilano, K., Lettieri-Barbato, D., et al., 2019. Nrf2 induction re-establishes a proper neuronal differentiation program in Friedreich's ataxia neural stem cells. Front. Cell. Neurosci. 13, 356.
Labuda, M., Labuda, D., Miranda, C., Poirier, J., Soong, B.W., Barucha, N.E., et al., 2000. Unique origin and specific ethnic distribution of the Friedreich's ataxia GAA expansion. Neurology 54, 2322–2324.
Lamarche, J.B., Cote, M., Lemieux, B., 1980. The cardiomyopathy of Friedreich's ataxia morphological observations in 3 cases. Can. J. Neurol. Sci. 7, 389–396.
Lefevre, S., Brossas, C., Auchere, F., Boggetto, N., Camadro, J.M., Santos, R., 2012. Apn1 AP-endonuclease is essential for the repair of oxidatively damaged DNA bases in yeast frataxin-deficient cells. Hum. Mol. Genet. 21, 4060–4072.
Lew, S.Y., Yow, Y.Y., Lim, L.W., Wong, K.H., 2020. Antioxidant-mediated protective role of *Hericiumerinaceus* (Bull.: Fr.) Pers. against oxidative damage in fibroblasts from Friedreich's ataxia patient. Food Sci. Technol. 40, 264–272.
Li, L., Voullaire, L., Sandi, C., Pook, M.A., Loannou, P.A., Delatycki, M.B., et al., 2013. Pharmacological screening using an FXN-EGFP cellular genomic reporter assay for the therapy of Friedreich ataxia. PLoS One 8, e55940.
Liu, Z., Ren, Z., Zhang, J., Chuang, C., Kandaswamy, E., Zhou, T., et al., 2018. Role of ROS and nutritional antioxidants in human diseases. Front. Physiol. 9, 477.
Llorens, J.V., Soriano, S., Calap-Quintana, P., Gonzalez-Cabo, P., Molto, M.D., 2019. The role of iron in Friedreich's ataxia: insights from studies in human tissues and cellular and animal models. Front. Neurosci. 13, 75.

Lodi, R., Hart, P.E., Rajagopalan, B., Taylor, D.J., Crilley, J.G., Bradley, J.L., et al., 2001. Antioxidant treatment improves in vivo cardiac and skeletal muscle bioenergetics in patients with Friedreich's ataxia. Ann. Neurol. 49, 590–596.
Lynch, D.R., Willi, S.M., Wilson, R.B., Cottiselli, M.G., Brigatti, K.W., Deutsch, E.C., et al., 2012. A0001 in Friedreich ataxia: biochemical characterization and effects in a clinical trial. Mov. Disord. 27, 1026–1033.
Marmolino, D., Manto, M., Acquaviva, F., Vergara, P., Ravella, A., Monticelli, A., et al., 2010. PGC-1alpha downregulation affects the antioxidant response in Friedreich's ataxia. PLoS One 5, e10025.
Martelli, A., Puccio, H., 2014. Dysregulation of cellular iron metabolism in Friedreich ataxia: from primary iron-sulfur cluster deficit to mitochondrial iron accumulation. Front. Pharmacol. 5, 130.
Michael, S., Petrocine, S.V., Qian, J., Lamarche, J.B., Knutson, M.D., Garrick, M.D., et al., 2006. Iron and iron-responsive proteins in the cardiomyopathy of Friedreich's ataxia. Cerebellum 5, 257–267.
Moreno-Ulloa, A., Nogueira, L., Rodriguez, A., Barboza, J., Hogan, M.C., Ceballos, G., et al., 2015. Recovery of indicators of mitochondrial biogenesis, oxidative stress, and aging with (−)-epicatechin in senile mice. J. Gerontol. A Biol. Sci. Med. Sci. 70, 1370–1378.
Orsucci, D., Mancuso, M., Ienco, E.C., LoGerfo, A., Siciliano, G., 2011. Targeting mitochondrial dysfunction and neurodegeneration by means of coenzyme Q10 and its analogues. Curr. Med. Chem. 18, 4053–4064.
Paupe, V., Dassa, E.P., Goncalves, S., Auchere, F., Lonn, M., Holmgren, A., et al., 2009. Impaired nuclear Nrf2 translocation undermines the oxidative stress response in Friedreich ataxia. PLoS One 4, e4253.
Petrillo, S., Piermarini, E., Pastore, A., Vasco, G., Schirinzi, T., Carrozzo, R., et al., 2017. Nrf2-inducers counteract neurodegeneration in Frataxin-silenced motor neurons: disclosing new therapeutic targets for Friedreich's ataxia. Int. J. Mol. Sci. 18, 2173.
Phang, M.W.L., Lew, S.Y., Chung, I., Lim, W.K., Lim, L.W., Wong, K.H., 2021. Therapeutic roles of natural remediesin combating hereditary ataxia: a systematic review. Chin. Med. 16, 15.
Piemonte, F., Pastore, A., Tozzi, G., Tagliacozzi, D., Santorelli, F.M., Carrozzo, R., et al., 2001. Glutathione in blood of patients with Friedreich's ataxia. Eur. J. Clin. Invest. 31, 1007–1111.
Probst, B.L., Trevino, I., McCauley, L., Bumeister, R., Dulubova, I., Wigley, W.C., 2015. RTA 408, a novel synthetic triterpenoid with broad anticancer and anti-inflammatory activity. PLoS One 10 (4), e0122942.
Puccio, H., Simon, D., Cossee, M., Criqui-Filipe, P., Tiziano, F., Melki, J., et al., 2001. Mouse models for Friedreich ataxia exhibit cardiomyopathy, sensory nerve defect and Fe-S enzyme deficiency followed by intramitochondrial iron deposits. Nat. Genet. 27, 181–186.
Ramirez, R.L., Qian, J., Santambrogio, P., Levi, S., Koeppen, A.H., 2012. Relation of cytosolic iron excess to cardiomyopathy of Friedreich's ataxia. Am. J. Cardiol. 110, 1820–1827.
Ramirez-Sanchez, I., Rodriguez, A., Moreno-Ulloa, A., Ceballos, G., Villarreal, F., 2016. (−)-Epicatechin-induced recovery of mitochondria from simulated diabetes: potential role of endothelial nitric oxide synthase. Diab. Vasc. Dis. Res. 13, 201–210.
Reata Pharmaceuticals, 2019. Reata Announces Positive Topline Results From the MOXIe Registrational Trial of Omaveloxolone in Patients With Friedreich's Ataxia. Globe News Wire, Los Angeles, CA.
Ren, J., Fan, C., Chen, N., Huang, J., Yang, Q., 2011. Resveratrol pretreatment attenuates cerebral ischemic injury by upregulating expression of transcription factor Nrf2 and HO-1 in rats. Neurochem. Res. 36, 2352–2362.
Rodriguez, L.R., Lapena, T., Calap-Quintana, P., Molto, M.D., Gonzalez-Cabo, P., Langa, J.A.N., 2020. Antioxidant therapies and oxidative stress in Friedreich's ataxia: the right path or just a diversion? Antioxidants 9, 664.

Rotig, A., de Lonlay, P., Chretien, D., Foury, F., Koenig, M., Sidi, D., et al., 1997. Aconitase and mitochondrial iron-sulphur protein deficiency in Friedreich ataxia. Nat. Genet. 17, 215–217.

Sanchez-Casis, G., Cote, M., Barbeau, A., 1976. Pathology of the heart in Friedreich's ataxia: review of the literature and report of one case. Can. J. Neurol. Sci. 3, 349–354.

Sano, M., Ernesto, C., Thomas, R.G., Klauber, M.R., Schafer, K., Grundman, M., et al., 1997. A controlled trial of selegiline, alpha-tocopherol, or both as treatment for Alzheimer's disease. N. Engl. J. Med. 336, 1216–1222.

Santos, R., Lefevre, S., Sliwa, D., Seguin, A., Camadro, J.M., Lesuisse, E., 2010. Friedreich ataxia: molecular mechanisms, redox considerations, and therapeutic opportunities. Antioxid. Redox Signal. 13, 651–690.

Sbodio, J.I., Snyder, S.H., Paul, B.D., 2019. Redox mechanisms in neurodegeneration: from disease outcomes to therapeutic opportunities. Antioxid. Redox Signal. 30, 1450–1499.

Schoenfeld, R.A., Napoli, E., Wong, A., Zhan, S., Reutenauer, L., Morin, D., et al., 2005. Frataxin deficiency alters heme pathway transcripts and decreases mitochondrial heme metabolites in mammalian cells. Hum. Mol. Genet. 14, 3787–3799.

Seco-Cervera, M., Gonzalez-Rodríguez, D., Ibáñez-Cabellos, J.S., Peiro-Chova, L., Gonzalez-Cabo, P., García-Lopez, E., et al., 2017. Circulating miR-323-3p is a biomarker for cardiomyopathy and an indicator of phenotypic variability in Friedreich's ataxia patients. Sci. Rep. 7, 5237.

Seco-Cervera, M., Gonzalez-Rodríguez, D., Ibanez-Cabellos, J.S., Peiro-Chova, L., Pallardó, F.V., Garcia-Gimenez, J.L., 2018. Small RNA-seq analysis of circulating miRNAs to identify phenotypic variability in Friedreich's ataxia patients. Sci. Data 5, 180021.

Sturm, B., Bistrich, U., Schranzhofer, M., Sarsero, J.P., Rauen, U., Scheiber-Mojdehkar, B., et al., 2005. Friedreich's ataxia, no changes in mitochondrial labile iron in human lymphoblasts and fibroblasts: a decrease in antioxidative capacity? J. Biol. Chem. 280, 6701–6708.

Takahashi, K., Tatsunami, R., Oba, T., Tampo, Y., 2010. Buthioninesulfoximine promotesmethylglyoxal-induced apoptotic cell death and oxidative stress in endothelial cells. Biol. Pharm. Bull. 33, 556–560.

Tan, G., Chen, L.S., Lonnerdal, B., Gellera, C., Taroni, F.A., Cortopassi, G.A., 2001. Frataxin expression rescues mitochondrial dysfunctions in FA cells. Hum. Mol. Genet. 10, 2099–2107.

Vang, O., Ahmad, N., Baile, C.A., Baur, J.A., Brown, K., Csiszar, A., et al., 2011. What is new for an old molecule? Systematic review and recommendations on the use of resveratrol. PLoS One 6, e19881.

Wilson, R.B., Roof, D.M., 1997. Respiratory deficiency due to loss of mitochondrial DNA in yeast lacking the frataxin homologue. Nat. Genet. 16, 352–357.

Wong, A., Yang, J., Cavadini, P., Gellera, C., Lonnerdal, B., Taroni, F., et al., 1999. The Friedreich's ataxia mutation confers cellular sensitivity to oxidant stress which is rescued by chelators of iron and calcium and inhibitors of apoptosis. Hum. Mol. Genet. 8, 425–430.

Worth, A.J., Basu, S.S., Deutsch, E.C., Hwang, W.T., Snyder, N.W., Lynch, D.R., et al., 2015. Stable isotopes and LC-MS for monitoring metabolic disturbances in Friedreich's ataxia platelets. Bioanalysis 7, 1843–1855.

Yiu, E.M., Tai, G., Peverill, R.E., Lee, K.J., Croft, K.D., Mori, T.A., et al., 2015. An open-label trial in Friedreich ataxia suggests clinical benefit with high-dose resveratrol, without effect on frataxin levels. J. Neurol. 262, 1344–1353.

Zaidi, S.M., Banu, N., 2004. Antioxidant potential of vitamins A, E and C in modulating oxidative stress in rat brain. Clin. Chim. Acta 340, 229–233.

Zuo, L., Hemmelgarn, B.T., Chuang, C.C., Best, T.M., 2015. The role of oxidative stress-induced epigenetic alterations in amyloid-beta production in Alzheimer's disease. Oxid. Med. Cell. Longev. 2015, 604658.

Further reading

Koeppen, A.H., Ramirez, R.L., Becker, A.B., Mazurkiewicz, J.E., 2016. Dorsal root ganglia in Friedreich ataxia: satellite cell proliferation and inflammation. Acta Neuropathol. Commun. 4, 46.

Retrotope Announces Phase I/II Clinical Trial Results of RT001 in Treatment of Friedreich's ataxia [press release] 2016. Available from: https://static1.squarespace.com/static/549af14ae4b004237f7bb71a/t/57dd7786f7e0ab741930775f/1474131955350/PR+Retrotope+Phase+1+Results.pdf.

CHAPTER 5

Plant extracts ameliorate Huntington's disease symptoms

Introduction

Huntington's disease (HD) is a neurodegenerative disease (ND) that affects 1–10 among every 100,000 people in western countries and is known as a minority disease. The contributory gene HTT encodes the Huntington protein with a hitherto unidentified role. The mutant Huntington induces a range of phenotypes such as oxidative stress and the activation of microglia and astrocytes, which result in chronic inflammation of the brain. Of the numerous targets of therapy, oxidative stress and inflammation have been adequately focused upon, and a few clinical trials have been initiated to target them (Bono-Yague et al., 2020).

Pathology

Free radicals

A few NDs like Alzheimer's disease (AD), Parkinson's disease (PD), amyotrophic lateral sclerosis (ALS), and HD share a common pathological link: accumulation of protein aggregates. Proteins that are inclined to aggregation form aggregates when the proteins disintegrate during the process of proteome dyshomeostasis. These systems include several cellular processes to mediate protein homeostasis during common incidents, and they are additionally provoked under stress conditions such as misfolded proteins, enhanced temperature, etc., thus augmenting oxidative stress (Balch et al., 2008). This stress is generated during an unusual increase in free radicals. Free radicals are generated due to the obstruction of cellular processes that influence mitochondrial function (Bhat et al., 2015). Oxidative stress is described as the impaired balance of redox homeostasis because of an unusual enhancement in free radicals and other reactive species that play a crucial role in cell signaling (Poli et al., 2004; Rhee, 2006). Free radicals are extremely reactive species that can induce oxidative damage to macromolecules such as DNA, lipids, and proteins. A few of the reactive

Plant Extracts in Neurodegenerative Diseases
https://doi.org/10.1016/B978-0-323-95762-5.00008-4

oxygen species (ROS) and nitrogen-derived reactive species (RNS), which are called RONS, can react with each other to generate other free radicals (Weidinger and Kozlov, 2015). Enhanced RONS in the mitochondria generate deleterious lipid peroxidation, which augments reactive lipid species (RLS), which are a resource of oxidative stress (Higdon et al., 2012; Hauck and Bernlohr, 2016). The best source of free radicals from ROS is organelles with an increased rate of oxygen use like mitochondria, endoplasmic reticulum, or peroxisomes (Trachootham et al., 2008; Fransen et al., 2012).

Antioxidants

The antioxidant system controls free radical generation to reestablish redox homeostasis. Natural cellular antioxidants include enzymes such as superoxide dismutase (SOD) (Fridovich, 1997), catalase (CAT) (Percy, 1984), glutathione system (GPx, GR, GST) (Aoyama and Nakaki, 2015), and thioredoxin system (Nordberg and Arner, 2001) and nonenzymatic molecules. Nonenzymatic antioxidant molecules can be exogenously fed to animal models and patients as drugs, despite their natural occurrence through diet such as vitamins (C and E), essential fatty acids (omega-3 and omega-6), carotenoids flavonoids, and trace metals (Se, Mn, Zn) (Halliwell, 2011; Ahmadinejad et al., 2017). A few antioxidants are endogenously synthesized by cell metabolism as defense against oxidative stress, like melatonin, coenzyme Q10, and reduced glutathione, among others (Beyer, 1992; Tan et al., 1993, 2000; Reiter et al., 2000; Van Bladeren, 2000; Turunen et al., 2004).

Oxidative stress in Huntington's disease

Healthy people commonly have 35 or fewer CAG repeats in HTT, whereas in those containing 36 and 39 repeats, deficient inducibility has been observed (Vonsattel and DiFiglia, 1998). Those who have ≥40 CAG repeats will certainly show evidence of the disease with full inducibility (Vonsattel and DiFiglia, 1998). Mutant CAG expansions encode unusually long polyQ tracks that generate impaired Htt with toxicity, which negatively influence the other nonrelated proteins, transcription factors, and protein quality organizing pathways (Soto, 2003). There is a counter connection between number of CAG repeats and age at HD onset (Illarioshkin et al., 1994; Rubinsztein et al., 1997). Despite the insufficient understanding of Htt, voluminous data has indicated its role in numerous cellular processes like modulation of energy metabolism, antiapoptotic activity, transcription regulation, DNA maintenance, axonal vesicle tracking, and cell signaling (Velier et al., 1998; Luthi-Carter, 2000; Ho, 2001; Zucker et al., 2005; Shirasaki et al., 2012).

Additionally, mammalian life is impossible without Htt, indicating that it plays an essential role in embryonic development (Reiner et al., 2003). The aggregation process is prompted by proteolysis of misfolded mHtt monomer to generate N-terminal fragments of expanded polyQ Htt (Lunkes et al., 2002; Arrasate and Finkbeiner, 2012). The N-terminal fragment of mHtt has a particular sequence (17 amino acids, N17) that serves as a nuclear import signal and facilitates aggregation formation in the nucleus despite the existence of aggregates in the cytoplasm (Lunkes et al., 2002). Furthermore, N17 encodes a particular domain that serves as an ROS sensor with a view to controlling the phosphorylation and location of Htt (DiGiovanni et al., 2016). Huntington is universally expressed in mammals, and this is quite applicable to the therapeutic intervention. Nevertheless, all tissues and cells are similarly affected (Saudou and Humbert, 2016). mHtt expression modulates the protein quality regulation pathways (autophagy and ubiquitin proteasome system), transcriptional regulators, and mitochondrial function (Martinez-Vicente et al., 2010; Hipp et al., 2012; Siddiqui et al., 2012; Seredenina and Luthi-Carter, 2012). Cytoplasmic mHtt aggregates bind to the mitochondrial membranes, because of their lipophilic nature, and attenuate complexes II and III from the electron transport chain (ETC), which subsequently mitigates intracellular ATP, perturbs the bioenergetics level, and enhances ROS (Jenkins et al., 1993; Brouillet et al., 1995; Gu et al., 1996; Bono-Yague et al., 2020).

Plant extracts

Plants and their parts with plenty of bioactive compounds have played an essential role in human health for years (Prasanth et al., 2019a). They can produce numerous bioactive compounds called "phytochemicals" that facilitate the protection of plants and show exemplary health benefits to humans (Nielsen et al., 2019). An array of medicinal properties offered by a plethora of plants include antioxidant, antiinflammatory, antitumor, and immunomodulatory effects (Rahimi et al., 2019), besides ameliorating cardiovascular ailments (Rouhi-Boroujeni et al., 2017), kidney stones (Akram and Idrees, 2019), and digestive diseases such as inflammatory bowel disease (Peng et al., 2019). The substantial benefit of plants and their medicinal properties is that they can be used for external applications or taken into the system in the form of food or beverage (Valduga et al., 2019; Prasanth et al., 2019b; Sharma et al., 2019).

Plant extracts show significant protection against HD in mice and rat models. Lycopene, a chief carotenoid available in tomato, acts as a potential antioxidant, and it improved mitochondrial function and alleviated locomotor and memory dysfunction in 3-nitropropionic acid (3-NP)-treated rats (Sies and Stahl, 1998; Sandhir et al., 2010). Similarly, oral administration of grape seed polyphenolic extract enhanced lifespan and ameliorated motor coordination in R6/2 mice (Wang et al., 2010). YAC128 mice treated with resveratrol also showed amelioration in their motor function and learning capability (Naia et al., 2017). Quercetin from flavonoid family with remarkable RONS scavenging efficacy improved mitochondrial function and motor performance and averted a reduction in antioxidant enzyme activity (Sandhir and Mehrotra, 2013). Additionally, a blend of quercetin and lycopene combated anxiety and depression in 3-NP-treated rats (Jain and Gangshettiwar, 2014). Sulforaphane, an isothiocyanate abundant in cruciferous vegetables, exhibited improvement in mitochondrial function in quinolinic acid-induced rats (Luis-Garcia et al., 2017). Rutin, a flavonoid found in buckwheat with significant free radical scavenging performance, demonstrated protection in 3-NP-induced rats. It enhanced the synthesis of glutathione (GSH), superoxide dismutase (SOD), and catalase (CAT), and decreased xanthine oxidase function, which plays a role in ROS production (Hanasaki et al., 1994; Kandemir et al., 2015; Kostic et al., 2015). Rutin ameliorated body weight loss, and reinstated the activity of antioxidant enzymes and motor and memory performance in 3-NP-induced rats (Suganya and Sumathi, 2017). Curcumin curtailed Htt aggregates in the striatum of CAG140 knock-in mice, though only moderate amelioration in motor performance was observed (Hickey et al., 2012). Elifani et al. (2019) reported that R6/2 mice supplemented with curcumin since birth improved motor function and activation of prosurvival pathways like ERK. However, these mice also showed a perturbance in gastrointestinal homeostasis, which provoked body weight loss. Conversely, curcumin treatment showed optimal gastrointestinal homeostasis indicating that further studies may result in robust output.

Clinical trials

The convincing results with α-tocopherol acquired in malonic acid-induced rats prompted their replication in a clinical study (Kalonia et al.,

2010). In line with this, 40 HD patients were given 3000 IU vitamin E/day and 33 patients were given sham. The study showed that there was no remarkable efficacy between the groups on both neurologic and neuropsychiatric symptoms. On the other hand, a therapeutic ability was observed on neurologic symptoms of patients in the early phases of HD (Peyser et al., 1995). Similarly, robust positive results of coenzyme Q10 on R6/2 mice prompted clinical studies (Ferrante et al., 2002). A phase I trial included 45 HD patients who were tested with various doses of 600, 1200, and 2400 mg daily, whereas 45 of other group was a control group. Despite the lack of functional improvement, the drug was suitably tolerated (Chandra et al., 2014). Consequently, they continued the trial with 240 treated HD patients and 240 controls in the 2CARE assay (Clinicaltrials.gov, number NCT00608881) (McGarry et al., 2017). Unfortunately, results could not show protective efficacy in patients with HD treated with 2400 mg coenzyme Q10/day.

Cleistocalyx nervosum var. *paniala*

Cleistocalyx nervosum var. *paniala*, a berry fruit found in Thailand, exerts several health benefits in vitro (Prasanth et al., 2019a). The soil nematode *Caenorhabditis elegans* is extensively used as a model to uncover a variety of parameters such as aging, development, reproduction, stress resistance, immune improvement, and pathology of NDs (Prasanth et al., 2016; Yunger et al., 2017; Sharika et al., 2018). *C. elegans* has been helpful to uncover various NDs such as AD (Yang et al., 2017), PD (Martinez et al., 2017), and dementia (Akinola, 2016). It has been considered a viable model due to feasibility in handling and maintenance, short life cycle and lifespan, and the presence of single gene mutants (Qian et al., 2015). It was the first eukaryotic organism with total uncovered sequence (The *C elegans* Sequencing Consortium, 1998). A wealth of studies on *C. elegans* treated with multifarious nutraceuticals isolated from plants unraveled the role of numerous genes and pathways that can mend lifespan and health (Dong et al., 2012). The initiation of SKN-1 by *C. nervosum* probably enhanced the lifespan and health in turn ameliorating stress resistance and neuroprotection (Hu et al., 2017). Several lines of evidence have shown that plant extracts and their bioactive compounds like *Paullinia cupana* (Boasquívis et al., 2018), rose essential oils (Zhu et al., 2017), and *Cratoxylum formosum* (Keowkase and Weerapreeyakul, 2016) showed neuroprotection through SKN-1-mediated antioxidant response. Nevertheless, these results are to be corroborated with molecular studies.

Guarana

Guarana, a well-known name of *Paullinia cupana* var. *sorbilis (Mart.)* seeds, has been extensively used in Brazil in traditional medicine, high energy drinks, and dietary components because of its initiative activity on the central nervous system (CNS) (Espinola et al., 1997; Kennedy et al., 2004; Otobone et al., 2007). The stimulant efficacy of guarana is linked to elevated caffeine quantity, a psychoactive pseudoalkaloid that plays a key role in human lifespan and aging-related neuropathologies (Kolahdouzan and Hamadeh, 2017; Boasquívis et al., 2018). Guarana also has methylxanthines, such as theophylline and theobromine, and polyphenols like catechins and epicatechins (da Silva Bittencourt et al., 2014; Yamaguti-Sasaki et al., 2007). These bioactive compounds induce antioxidant (Bittencourt et al., 2013; de Portella et al., 2013; Yonekura et al., 2016), antimicrobial (Yamaguti-Sasaki et al., 2007), and chemoprophylactic activities in carcinogenesis (Fukumasu et al., 2006, 2011) and antigenotoxic effects (Leite et al., 2011).

Protective efficacy of guarana against neurodegenerative diseases

Caffeine and coffee extract curtailed protein aggregation and toxicity in both HD and AD models of *C. elegans* (Dostal et al., 2010; Sutphin et al., 2012; Brunquell et al., 2018). Guarana water extracts decreased the formation of polyglutamine (polyQ) aggregates observed in *C. elegans* muscle, indicating a protective efficacy of guarana in HD (Peixoto et al., 2017). Although studies have demonstrated the neuroprotective efficacy of guarana, there is scant evidence on its mechanism of action. Therefore, *Caenorhabditis elegans* has been chosen as a viable model to unravel the molecular underpinnings that regulate stress responses, aging, and NDs. Numerous aggregation-prone human proteins linked to NDs have been studied in a wide range of *C. elegans* tissues to have an insight into protein misfolding and aggregation (Dimitriadi and Hart, 2010; Li and Le, 2013; Lublin and Link, 2013). HSF-1 knockdown enhanced polyQ protein accumulation in *C. elegans* models of HD (Nollen et al., 2004). Similarly, silencing age-1 to decrease protein aggregates in a DAF-16-dependent manner exhibited an association between aging-linked pathways and polyQ accumulation (Morley et al., 2002). Guarana hydroalcoholic extract showed exemplary protective efficacy in both AD and HD models by deferred or decreased toxic phenotypes correlated with protein misfolding and accumulation (Boasquívis et al., 2018).

Studies using green fluorescent protein (GFP) unraveled that guarana hydroalcoholic extract (GHE) treatment enhanced the expression of proteins linked to protein homeostasis, lysosome degradation, and oxidative stress responses. Biochemical analysis demonstrated that GHE enhances proteasome activity and decreases intracellular ROS (Boasquívis et al., 2018).

Conclusions and future perspectives

NDs such as AD, PD, and HD share common pathological events such as oxidative stress and deposition of protein aggregates. Therefore, antioxidants and bioactive molecules such as curcumin, which can inhibit ROS generation and protein aggregation, have been used as appropriate therapeutics. Several traditional medicines such as Indian and Chinese traditional medicines comprise bioactive compounds that significantly ameliorate several CNS-related diseases. They have also been used for chronic diseases for years. Despite the remarkable studies to overcome HD, it has been a herculean task to draw a bead on its therapeutic compounds. Although voluminous data showed decrease of phenotypes linked to oxidative stress, the results could not be replicated in animal models and patients. A plethora of bioactive compounds have been isolated and the protective efficacy against HD has been studied. However, a few clinical trials did not show appropriate therapeutic efficacy against HD symptoms despite the significant therapeutic efficacy in cellular and animal models of HD. Therefore, there is a growing need to focus on several more novel molecules that can enhance therapeutic efficacy against HD. In addition, a thorough understanding into the existing molecules and the novel way of their use is warranted. The bioactive molecules with multiple targets will be significant to develop potential armamentarium. The significant understanding of the molecular underpinnings entailed in the pathology of HD is also of urgent need since it is the basis for designing appropriate therapeutics.

References

Ahmadinejad, F., Geir Moller, S., Hashemzadeh-Chaleshtori, M., Bidkhori, G., Jami, M.S., 2017. Molecular mechanisms behind free radical scavengers function against oxidative stress. Antioxidants 6, 51.

Akinola, O.B., 2016. Sweet old memories: a review of the experimental models of the association between diabetes, senility and dementia. Metab. Brain Dis. 31 (5), 1003–1010.

Akram, M., Idrees, M., 2019. Progress and prospects in the management of kidney stones and developments in phytotherapeutic modalities. Int. J. Immunopathol. Pharmacol. 33, 1–5.

Aoyama, K., Nakaki, T., 2015. Glutathione in cellular redox homeostasis: association with the excitatory amino acid carrier 1 (EAAC1). Molecules 20, 8742–8758.

Arrasate, M., Finkbeiner, S., 2012. Protein aggregates in Huntington's disease. Exp. Neurol. 238, 1–11.

Balch, W.E., Morimoto, R.I., Dillin, A., Kelly, J.W., 2008. Adapting proteostasis for disease intervention. Science 319, 916–919.

Beyer, R.E., 1992. An analysis of the role of coenzyme Q in free radical generation and as an antioxidant. Biochem. Cell Biol. 70, 390–403.

Bhat, A.H., Dar, K.B., Anees, S., Zargar, M.A., Masood, A., Sofi, M.A., et al., 2015. Oxidative stress, mitochondrial dysfunction and neurodegenerative diseases; a mechanistic insight. Biomed. Pharmacother. 74, 101–110.

Bittencourt, L.S., Machado, D.C., Machado, M.M., Santos Dos, G.F.F., Algarv, T.D., Marinowic, D.R., et al., 2013. The protective effects of guaraná extract (Paullinia cupana) on fibroblast NIH-3T3 cells exposed to sodium nitroprusside. Food Chem. Toxicol. 53, 119–125.

Boasquívis, P.F., Silva, G.M.M., Paiva, F.A., Cavalcanti, R.M., Nunez, C.V., de Paula Oliveira, R., 2018. Guarana (*Paullinia cupana*) extract protects *Caenorhabditis elegans* models for Alzheimer disease and Huntington disease through activation of antioxidant and protein degradation pathways. Oxid. Med. Cell. Longev. 2018, 9241308. 16 pp.

Bono-Yague, J., Gomez-Escribano, A.P., Millan, J.M., Vazquez-Manrique, R.P., 2020. Reactive species in Huntington disease: are they really the radicals you want to catch? Antioxidants 9, 577.

Brouillet, E., Hantraye, P., Ferrante, R.J., Dolan, R., Leroy-Willig, A., Kowall, N.W., et al., 1995. Chronic mitochondrial energy impairment produces selective striatal degeneration and abnormal choreiform movements in primates. Proc. Natl. Acad. Sci. U. S. A. 92, 7105–7109.

Brunquell, J., Morris, S., Snyder, A., Westerheide, S.D., 2018. Coffee extract and caffeine enhance the heat shock response and promote proteostasis in an HSF-1-dependent manner in Caenorhabditis elegans. Cell Stress Chaperones 23 (1), 65–75.

Chandra, A., Johri, A., Beal, M.F., 2014. Prospects for neuroprotective therapies in prodromal Huntington's disease. Mov. Disord. 29, 285–293.

da Silva Bittencourt, L., Zeidán-Chuliá, F., Yatsu, F.K.J., Schnorr, C.E., Moresco, C.S., et al., 2014. Guarana (Paullinia cupana Mart.) prevents β-amyloid aggregation, generation of advanced glycation-end products (AGEs), and acrolein-induced cytotoxicity on human neuronal-like cells. Phytother. Res. 28 (11), 1615–1624.

de Portella, R.L., Barcelos, R.P., da Rosa, E.J.F., Ribeiro, E.E., da Cruz, I.B.M., Suleiman, L., et al., 2013. Guaraná (Paullinia cupana Kunth) effects on LDL oxidation in elderly people: an in vitro and in vivo study. Lipids Health Dis. 12 (1), 12.

DiGiovanni, L.F., Mocle, A.J., Xia, J., Truant, R., 2016. Huntingtin N17 domain is a reactive oxygen species sensor regulating huntingtin phosphorylation and localization. Hum. Mol. Genet. 25, 3937–3945.

Dimitriadi, M., Hart, A.C., 2010. Neurodegenerative disorders: insights from the nematode Caenorhabditis elegans. Neurobiol. Dis. 40 (1), 4–11.

Dong, Y., Guha, S., Sun, X., Cao, M., Wang, X., Zou, S., 2012. Nutraceutical interventions for promoting healthy aging in invertebrate models. Oxid. Med. Cell. Longev., 718491. 10 pp.

Dostal, V., Roberts, C.M., Link, C.D., 2010. Genetic mechanisms of coffee extract protection in a Caenorhabditis elegans model of β-amyloid peptide toxicity. Genetics 186 (3), 857–866.

Elifani, F., Amico, E., Pepe, G., Capocci, L., Castaldo, S., Rosa, P., et al., 2019. Curcumin dietary supplementation ameliorates disease phenotype in an animal model of Huntington's disease. Hum. Mol. Genet. 28, 4012–4021.

Espinola, E.B., Dias, R.F., Mattei, R., Carlini, E.A., 1997. Pharmacological activity of guarana (Paullinia cupana Mart.) in laboratory animals. J. Ethnopharmacol. 55 (3), 223–229.

Ferrante, R.J., Andreassen, O.A., Dedeoglu, A., Ferrante, K.L., Jenkins, B.G., Hersch, S.M., et al., 2002. Therapeutic effects of coenzyme Q10 and remacemide in transgenic mouse models of Huntington's disease. J. Neurosci. 22, 1592–1599.

Fransen, M., Nordgren, M., Wang, B., Apanasets, O., 2012. Role of peroxisomes in ROS/RNS-metabolism: implications for human disease. Biochim. Biophys. Acta Mol. Basis Dis. 1822, 1363–1373.

Fridovich, I., 1997. Superoxide anion radical ($O_2^{\bullet-}$), superoxide dismutases, and related matters. J. Biol. Chem. 272, 18515–18517.

Fukumasu, H., da Silva, T.C., Avanzo, J.L., de Lima, C.E., Mackowiak, I.I., Atroch, A., et al., 2006. Chemopreventive effects of Paullinia cupana Mart var. sorbilis, the guaraná, on mouse hepatocarcinogenesis. Cancer Lett. 233 (1), 158–164.

Fukumasu, H., Latorre, A.O., Zaidan-Dagli, M.L., 2011. Paullinia cupana Mart. var. sorbilis, guarana, increases survival of Ehrlich ascites carcinoma (EAC) bearing mice by decreasing cyclin-D1 expression and inducing a G0/G1 cell cycle arrest in EAC cells. Phytother. Res. 25 (1), 11–16.

Gu, M., Gash, M.T., Mann, V.M., Javoy-Agid, F., Cooper, J.M., Schapira, A.H.V., 1996. Mitochondrial defect in Huntington's disease caudate nucleus. Ann. Neurol. 39, 385–389.

Halliwell, B., 2011. Free radicals and antioxidants–quo vadis? Trends Pharmacol. Sci. 32, 125–130.

Hanasaki, Y., Ogawa, S., Fukui, S., 1994. The correlation between active oxygens scavenging and antioxidative effects of flavonoids. Free Radic. Biol. Med. 16, 845–850.

Hauck, A.K., Bernlohr, D.A., 2016. Oxidative stress and lipotoxicity. J. Lipid Res. 57, 1976–1986.

Hickey, M.A., Zhu, C., Medvedeva, V., Lerner, R.P., Patassini, S., Franich, N.R., et al., 2012. Improvement of neuropathology and transcriptional deficits in CAG 140 knock-in mice supports a beneficial effect of dietary curcumin in Huntington's disease. Mol. Neurodegener. 7, 12.

Higdon, A., Diers, A.R., Oh, J.Y., Landar, A., Darley-Usmar, V.M., 2012. Cell signalling by reactive lipid species: new concepts and molecular mechanisms. Biochem. J. 442, 453–464.

Hipp, M.S., Patel, C.N., Bersuker, K., Riley, B.E., Kaiser, S.E., Shaler, T.A., et al., 2012. Indirect inhibition of 26S proteasome activity in a cellular model of Huntington's disease. J. Cell Biol. 196, 573–587.

Ho, L.W., 2001. Wild type huntingtin reduces the cellular toxicity of mutant huntingtinin mammalian cell models of Huntington's disease. J. Med. Genet. 38, 450–452.

Hu, Q., Amora, D.R.D., MacNeil, L.T., Walhout, A.J.M., Kubiseski, T.J., 2017. The oxidative stress response in Caenorhabditis elegans requires the GATA transcription factor ELT-3 and SKN-1/Nrf2. Genetics 206 (4), 1909–1922.

Illarioshkin, S.N., Igarashi, S., Onodera, O., Markova, E.D., Nikolskaya, N.N., Tanaka, H., et al., 1994. Trinucleotide repeat length and rate of progression of Huntington's disease. Ann. Neurol. 36, 630–635.

Jain, D., Gangshettiwar, A., 2014. Combination of lycopene, quercetin and poloxamer188 alleviates anxiety and depression in 3-nitropropionic acid-induced Huntingtons disease in rats. J. Intercult. Ethnopharmacol. 3, 186–191.

Jenkins, B.G., Koroshetz, W.J., Beal, M.F., Rosen, B.R., 1993. Evidence for impairment of energy metabolism in vivo in Huntington's disease using localized 1H NMR spectroscopy. Neurology 43, 2689.

Kalonia, H., Kumar, P., Kumar, A., 2010. Targeting oxidative stress attenuates malonic acid induced Huntington like behavioral and mitochondrial alterations in rats. Eur. J. Pharmacol. 634, 46–52.

Kandemir, F.M., Ozkaraca, M., Yildirim, B.A., Hanedan, B., Kirbas, A., Kilic, K., et al., 2015. Rutin attenuates gentamicin-induced renal damage by reducing oxidative stress, inflammation, apoptosis and autophagy in rats. Ren. Fail. 37, 518–525.

Kennedy, D.O., Haskell, C.F., Wesnes, K.A., Scholey, A.B., 2004. Improved cognitive performance in human volunteers following administration of guarana (Paullinia cupana) extract: comparison and interaction with Panax ginseng. Pharmacol. Biochem. Behav. 79 (3), 401–411.

Keowkase, R., Weerapreeyakul, N., 2016. Cratoxylum formosum extract protects against amyloid-beta toxicity in a Caenorhabditis elegans model of Alzheimer's disease. Planta Med. 82 (6), 516–523.

Kolahdouzan, M., Hamadeh, M.J., 2017. The neuroprotective effects of caffeine in neurodegenerative diseases. CNS Neurosci. Ther. 23 (4), 272–290.

Kostic, D., Dimitrijevic, D., Stojanovic, G., Palic, I., Dordevic, A., Ickovski, J., 2015. Xanthine oxidase: isolation, assays of activity, and inhibition. J. Chem. 2015, 294858.

Leite, R.P., Wada, R.S., Monteiro, J.C., Predes, F.S., Dolder, H., 2011. Protectivee of Guarana (*Paullinia cupana* var. sorbilis) pre-treatment on cadmium-induced damages in adult Wistar testis. Biol. Trace Elem. Res. 141 (1–3), 262–274.

Li, J., Le, W., 2013. Modeling neurodegenerative diseases in Caenorhabditis elegans. Exp. Neurol. 250, 94–103.

Lublin, A.L., Link, C.D., 2013. Alzheimer's disease drug discovery: in vivo screening using *Caenorhabditis elegans* as a model for β-amyloid peptide-induced toxicity. Drug Discov. Today Technol. 10 (1), e115–e119.

Luis-Garcia, E.R., Limon-Pacheco, J.H., Serrano-Garcia, N., Hernandez-Perez, A.D., Pedraza-Chaverri, J., Orozco-Ibarra, M., 2017. Sulforaphane prevents quinolinic acid-induced mitochondrial dysfunction in rat striatum. J. Biochem. Mol. Toxicol. 31, e21837.

Lunkes, A., Lindenberg, K.S., Ben-Haiem, L., Weber, C., Devys, D., Wehrmeyer, G.B., et al., 2002. Proteases acting on mutant huntingtin generate cleaved products that differentially build up cytoplasmic and nuclear inclusions. Mol. Cell 10, 259–269.

Luthi-Carter, R., 2000. Decreased expression of striatal signaling genes in a mouse model of Huntington's disease. Hum. Mol. Genet. 9, 1259–1271.

Martinez, B.A., Caldwell, K.A., Caldwell, G.A., 2017. C. elegans as a model system to accelerate discovery for Parkinson's disease. Curr. Opin. Genet. Dev. 44, 102–109.

Martinez-Vicente, M., Talloczy, Z., Wong, E., Tang, G., Koga, H., Kaushik, S., et al., 2010. Cargo recognition failure is responsible for inefficient autophagy in Huntington's disease. Nat. Neurosci. 13, 567–576.

McGarry, A., McDermott, M., Kieburtz, K., de Blieck, E.A., Beal, F., Marder, K., et al., 2017. A randomized, double-blind, placebo-controlled trial of coenzyme Q10 in Huntington disease. Neurology 88, 152–159.

Morley, J.F., Brignull, H.R., Weyers, J.J., Morimoto, R.I., 2002. The threshold for polyglutamine-expansion protein aggregation and cellular toxicity is dynamic and influenced by aging in Caenorhabditis elegans. Proc. Natl. Acad. Sci. U.S.A. 99 (16), 10417–10422.

Naia, L., Rosenstock, T.R., Oliveira, A.M., Oliveira-Sousa, S.I., Caldeira, G.L., Carmo, C., et al., 2017. Comparative mitochondrial-based protective effects of resveratrol and nicotinamide in Huntington's disease models. Mol. Neurobiol. 54, 5385–5399.

Nielsen, E., Temporiti, M.E.E., Cella, R., 2019. Improvement of phytochemical production by plant cells and organ culture and by genetic engineering. Plant Cell Rep. 38 (10), 1199–1215.

Nollen, E.A.A., Garcia, S.M., van Haaften, G., Kim, S., Chavez, A., Morimoto, R.I., et al., 2004. Genome wide RNA interference screen identifies previously undescribed regulators of polyglutamine aggregation. Proc. Natl. Acad. Sci. U. S. A. 101 (17), 6403–6408. 2004.

Nordberg, J., Arner, E.S.J., 2001. Reactive oxygen species, antioxidants, and the mammalian thioredoxin system. Free Radic. Biol. Med. 31, 1287–1312.

Otobone, F.J., Sanches, A.C.C., Nagae, R., Martins, J.V.C., Sela, V.R., de Mello, J.C.P., et al., 2007. Effect of lyophilized extracts from guarana seeds [*Paullinia cupana var. sorbilis (Mart.) Ducke*] on behavioral profiles in rats. Phytother. Res. 21 (6), 531–535.

Peixoto, H., Roxo, M., Rohrig, T., Richling, E., Wang, X., Wink, M., 2017. Anti-aging and antioxidant potential of Paullinia cupana var. sorbilis: findings in Caenorhabditis elegans indicate a new utilization for roasted seeds of guarana. Medicines 4 (3), 61.

Peng, J., Zheng, T.T., Li, X., Liang, Y., Wang, L., Huang, Y., et al., 2019. Plant-derived alkaloids: the promising disease-modifying agents for inflammatory bowel disease. Front. Pharmacol. 10, 351.

Percy, M.E., 1984. Catalase: an old enzyme with a new role? Can. J. Biochem. Cell Biol. 62, 1006–1014.

Peyser, C.E., Folstein, M., Chase, G.A., Starkstein, S., Brandt, J., Cockrell, J.R., et al., 1995. Trial of d-alpha-tocopherol in Huntington's disease. Am. J. Psychiatry 152, 1771–1775.

Poli, G., Leonarduzzi, G., Biasi, F., Chiarpotto, E., 2004. Oxidative stress and cell signalling. Curr. Med. Chem. 11, 1163–1182.

Prasanth, M.I., Santoshram, G.S., Bhaskar, J.P., Balamurugan, K., 2016. Ultraviolet-A triggers photoaging in model nematode Caenorhabditis elegans in a DAF-16 dependent pathway. Age 38 (1), 27.

Prasanth, M.I., Brimson, J.M., Chuchawankul, S., Sukprasansap, M., Tencomnao, T., 2019a. Antiaging, stress resistance, and neuroprotective efficacies of Cleistocalyx nervosum var. paniala fruit extracts using *Caenorhabditis elegans* model. Oxid. Med. Cell. Longev. 2019, 1–14.

Prasanth, M.I., Sivamaruthi, B.S., Chaiyasut, C., Tencomnao, T., 2019b. A review of the role of green tea (Camellia sinensis) in antiphotoaging, stress resistance, neuroprotection and autophagy. Nutrients 11 (2), 474.

Qian, H., Xu, X., Niklason, L.E., 2015. PCH-2 regulates Caenorhabditis elegans lifespan. Aging 7 (1), 1–13.

Rahimi, V.B., Ajam, F., Rakhshandeh, H., Askari, V.R., 2019. A pharmacological review on Portulaca oleracea L.: focusing on anti-inflammatory, anti-oxidant, immuno-modulatory and antitumor activities. J. Pharmacopuncture 22 (1), 7–15.

Reiner, A., Dragatsis, I., Zeitlin, S., Goldowitz, D., 2003. Wild-type huntingtin plays a role in brain development and neuronal survival. Mol. Neurobiol. 28, 259–276.

Reiter, R.J., Tan, D.X., Acuna-Castroviejo, D., Burkhardt, S., Karbownik, M., 2000. Melatonin: mechanisms and actions as an antioxidant. Curr. Top. Biophys. 24, 171–184.

Rhee, S.G., 2006. Cell signalling: H_2O_2, a necessary evil for cell signalling. Science 312, 1882–1883.

Rouhi-Boroujeni, H., Heidarian, E., Rouhi-Boroujeni, H., Deris, F., Rafieian-Kopaei, M., 2017. Medicinal plants with multiple effects on cardiovascular diseases: a systematic review. Curr. Pharm. Des. 23 (7), 999–1015.

Rubinsztein, D.C., Leggo, J., Chiano, M., Dodge, A., Norbury, G., Rosser, E., et al., 1997. Genotypes at the GluR6 kainate receptor locus are associated with variation in the age of onset of Huntington disease. Proc. Natl. Acad. Sci. U. S. A. 94, 3872–3876.

Sandhir, R., Mehrotra, A., 2013. Quercetin supplementation is effective in improving mitochondrial dysfunctions induced by 3-nitropropionic acid: implications in Huntington's disease. Biochim. Biophys. Acta Mol. Basis Dis. 1832, 421–430.

Sandhir, R., Mehrotra, A., Kamboj, S.S., 2010. Lycopene prevents 3-nitropropionic acid-induced mitochondrial oxidative stress and dysfunctions in nervous system. Neurochem. Int. 57, 579–587.

Saudou, F., Humbert, S., 2016. The biology of huntingtin. Neuron 89, 910–926.

Seredenina, T., Luthi-Carter, R., 2012. What have we learned from gene expression profiles in Huntington's disease? Neurobiol. Dis. 45, 83–98.

Sharika, R., Subbaiah, P., Balamurugan, K., 2018. Studies on reproductive stress caused by candidate Gram positive and Gram negative bacteria using model organism, Caenorhabditis elegans. Gene 649, 113–126.

Sharma, R., Martins, N., Kuca, K., Chaudhary, A., Kabra, A., Rao, M.M., et al., 2019. Chyawanprash: a traditional Indian bioactive health supplement. Biomol. Ther. 9 (5), 161.

Shirasaki, D.I., Greiner, E.R., Al-Ramahi, I., Gray, M., Boontheung, P., Geschwind, D.H., et al., 2012. Network Organization of the Huntingtin proteomic interactome in mammalian brain. Neuron 75, 41–57.

Siddiqui, A., Rivera-Sánchez, S., Castro, M.D.R., Acevedo-Torres, K., Rane, A., Torres-Ramos, C.A., et al., 2012. Mitochondrial DNA damage is associated with reduced mitochondrial bioenergetics in Huntington's disease. Free Radic. Biol. Med. 53, 1478–1488.

Sies, H., Stahl, W., 1998. Lycopene: antioxidant and biological effects and its bioavailability in the human. Proc. Soc. Exp. Biol. Med. 218, 121–124.

Soto, C., 2003. Unfolding the role of protein misfolding in neurodegenerative diseases. Nat. Rev. Neurosci. 4, 49–60.

Suganya, S.N., Sumathi, T., 2017. Effect of rutin against a mitochondrial toxin, 3-nitropropionicacid induced biochemical, behavioral and histological alterations—a pilot study on Huntington's disease model in rats. Metab. Brain Dis. 32, 471–481.

Sutphin, G.L., Bishop, E., Yanos, M.E., Moller, R.M., Kaeberlein, M., 2012. Caffeine extends life span, improves health span, and delays age-associated pathology in Caenorhabditis elegans. Longev. Healthspan 1 (1), 9.

Tan, D.X., Chen, L., Poeggeler, B., Manchester, L., Reiter, R., 1993. Melatonin: a potent, endogenous hydroxyl radical scavenger. Endocr. J. 1, 57–60.

Tan, D.X., Manchester, L.C., Reiter, R.J., Qi, W.B., Karbownik, M., Calvo, J.R., 2000. Significance of melatonin in antioxidative defense system: reactions and products. Neurosignals 9, 137–159.

The C elegans Sequencing Consortium, 1998. Genome sequence of the nematode C. elegans: a platform for investigating biology. Science 282 (5396), 2012–2018.

Trachootham, D., Lu, W., Ogasawara, M.A., Valle, N.R.D., Huang, P., 2008. Redox regulation of cell survival. Antioxid. Redox Signal. 10, 1343–1374.

Turunen, M., Olsson, J., Dallner, G., 2004. Metabolism and function of coenzyme Q. Biochim. Biophys. Acta 1660, 171–199.

Valduga, A.T., Gonçalves, I.L., Magri, E., Delalibera Finzer, J.R., 2019. Chemistry, pharmacology and new trends in traditional functional and medicinal beverages. Food Res. Int. 120, 478–503.

Van Bladeren, P.J., 2000. Glutathione conjugation as a bioactivation reaction. Chem. Biol. Interact. 129, 61–76.

Velier, J., Kim, M., Schwarz, C., Kim, T.W., Sapp, E., Chase, K., et al., 1998. Wild-type and mutant huntingtins function in vesicle trafficking in the secretory and endocytic pathways. Exp. Neurol. 152, 34–40.

Vonsattel, J.P.G., DiFiglia, M., 1998. Huntington disease. J. Neuropathol. Exp. Neurol. 57, 369–384.

Wang, J., Pfleger, C., Friedman, L., Vittorino, R., Zhao, W., Qian, X., et al., 2010. Potential application of grape derived polyphenols in Huntington's disease. Transl. Neurosci. 1, 95–100.

Weidinger, A., Kozlov, A., 2015. Biological activities of reactive oxygen and nitrogen species: oxidative stress versus signal transduction. Biomol. Ther. 5, 472–484.

Yamaguti-Sasaki, E., Ito, L., Canteli, V., Ushirobira, T.M.A., Ueda-Nakamura, T., Dias Filho, B.P., et al., 2007. Antioxidant capacity and in vitro prevention of dental plaque formation by extracts and condensed tannins of Paullinia cupana. Molecules 12 (8), 1950–1963.

Yang, J., Huang, X.B., Wan, Q.L., Ding, A., Yang, Z., Qui, M., et al., 2017. Otophylloside B protects against Aβ toxicity in Caenorhabditis elegans models of Alzheimer's disease. Nat. Prod. Bioprospect 7 (2), 207–214.

Yonekura, L., Martins, C.A., Sampaio, G.R., Monteiro, M.P., Cesar, L.A.M., Mioto, B.M., et al., 2016. Bioavailability of catechins from guaraná (Paullinia cupana) and its effect on antioxidant enzymes and other oxidative stress markers in healthy human subjects. Food Funct. 7 (7), 2970–2978.

Yunger, E., Safra, M., Levi-Ferber, M., Haviv-Chesner, A., Henis-Korenblit, S., 2017. Innate immunity mediated longevity and longevity induced by germ cell removal converge on the C-type lectin domain protein IRG-7. PLoS Genet. 13 (2), e1006577.

Zhu, S., Li, H., Dong, J., Yang, W., Liu, T., Wang, Y., et al., 2017. Rose essential oil delayed Alzheimer's disease-like symptoms by SKN-1 pathway in C. elegans. J. Agric. Food Chem. 65 (40), 8855–8865.

Zucker, B., Luthi-Carter, R., Kama, J.A., Dunah, A.W., Stern, E.A., Fox, J.H., et al., 2005. Transcriptional dysregulation in striatal projection- and interneurons in a mouse model of Huntington's disease: neuronal selectivity and potential neuroprotective role of HAP1. Hum. Mol. Genet. 14, 179–189.

CHAPTER 6

Phytochemicals and their potential protective effects against spinocerebellar ataxia

Introduction

Pathology

Ataxia is described as the perturbance in the regulation of bodily movements that can arise sporadically or are inherited or acquired (Akbar and Ashizawa, 2015; Phang et al., 2021). The pathological aspects of spinocerebellar ataxia (SCA) include neurodegenerative effects such as atrophy in the cerebellum affecting the brain stem and spinal cord (Schols et al., 2004; Klockgether et al., 2019; Khan et al., 2020). All varieties of SCA are linked to the lack of synchronization in eye and hand movements including speech. SCA generally has adult onset, and the development is steady exacerbating with time (Groth and Berman, 2018). Currently, 44 types of SCA have been diagnosed and categorized from SCA1 to SCA44. Several SCAs are the result of CAG nucleotide repeat expansion encoding polyglutamine (polyQ), leading to the contribution of toxic polyQ protein (Sullivan et al., 2018; Khan et al., 2020).

SCA type 17 (SCA 17), a polyQ disease, is triggered due to the expansion of CAG/CAA repeats (43 repeats) in the TATA box-binding protein (TBP) gene, resulting in protein aggregation in patients with SCA 17 (Fujigasaki et al., 2001; Nakamura et al., 2001; Zuhlke et al., 2001; Huang et al., 2016). Pathological nonnative protein aggregation, impaired cellular heat shock response (HSR), and perturbed protein quality regulation pathways are the underlying causes of a wide range of neurodegenerative diseases (NDs) (Auluck and Bonini, 2002; Neef et al., 2010; Maheshwari et al., 2014; Labbadia and Morimoto, 2015; Nelson et al., 2016).

Perturbed calcium signaling probably initiates protein aggregation and neuronal death in polyQ diseases. Therefore, the attenuation of calcium overfilling by a therapeutic may suffice the needs for SCA 17 treatment (Huang et al., 2016). The sodium dodecyl sulfate (SDS)-resistant

https://doi.org/10.1016/B978-0-323-95762-5.00004-7

aggregation of insoluble proteins is enhanced after the glutamate-instigated inflow of calcium, and the aggregation is decreased by the suppression of calcium influx in a model of SCA 3 (Koch et al., 2011; Huang et al., 2016). Additionally, the activation of calcium-dependent proteolytic calpains exacerbates SCA 3 pathology by enhancing protein aggregation; nevertheless, calpastatin, an inhibitor of calpains, mitigates aggregation (Simoes et al., 2012; Hubener et al., 2013; Huang et al., 2016).

Hereditary ataxia (HA) denotes a group of genetically heterogeneous NDs occurring due to the impairment of the cerebellum or disturbance of association between the cerebellum and other areas of the central nervous system (Albernaz et al., 2019). The characteristics of HA include seizure, pyramidal syndrome, peripheral neuropathy, movement disorder, and cognitive impairment (Teive and Ashizawa, 2015; Phang et al., 2021). In addition, HA patients also experience instability of stance and gait, nystagmus, dysarthria, dysmetria, and clumsiness (Sarva and Shanker, 2014; Phang et al., 2021).

Dendrite

The aggregation of misfolded polyQ proteins induced by Q extension exhibits protein toxicity, which leads to neuropathic characteristics like dendrite deficiency, resulting in neuronal dysfunction and neuronal cell death (Morley et al., 2002; Bates, 2003; Kwon et al., 2018). Growing lines of evidence have shown that abnormal dendritic morphologies are associated with impaired brain functions like memory storage and cognition (Kulkarni and Firestein, 2012; Stuart and Spruston, 2015). Since dendritic pathology is common in patients suffering from polyQ diseases, it is essential to investigate the role of misfolded polyQ proteins in dendrite pathology (Graveland et al., 1985; Sotrel et al., 1993). Dendrite deficiency was linked to an impaired actin cytoskeletal structure and weakened subcellular distribution of Golgi outposts in *Drosophila* models of SCA2, which is also called Machado-Joseph disease (MJD) (Lee et al., 2011; Chung et al., 2017). Coiled-coil structures of polyQ proteins play a vital role in neuronal toxicity, which is initiated by coiled-coil-to-coiled-coil communications with the nuclear targets of proteins (Kwon et al., 2018).

Plant extracts

Medicinal plants contain a wealth of bioactive compounds that have been used to cure many types of diseases for many years. These plants remain as a source for traditional medicine, and approximately 3.3 billion people from underdeveloped countries take them to cure illnesses (Davidson-Hunt, 2000; Jeyasri et al., 2020). Natural remedies play a pivotal role as therapeutic agents for multifarious NDs (Pohl and Kong Thoo Lin, 2018; Ayaz et al., 2019). They are a substantial source for the development of drugs or agents with significant therapeutic efficacy. Holistic practitioners employ natural remedies due to their good reception and safety compared to synthetic drugs. The primary potential property of natural remedies is their affinity to the target protein or specific biomolecules in living organisms (Ayaz et al., 2019). Natural metabolites with higher structural multiplicity, bioactivity, and intricacy are suitable as substrates for a few of the transporter systems and are localized in the targeted intracellular site of action (Harvey et al., 2015).

Curcumin (Obulesu, 2021), resveratrol (Dendi and Obulesu, 2020), quercetin (Obulesu, 2020), green tea catechins (Khaleel and Obulesu, 2020), and salicylate are a few of the extensively studied biological compounds with understood copious molecular targets and mechanisms of action (Koeberle and Werz, 2014). Nevertheless, a disadvantage of biological compounds with numerous targets is their affinity to more proteins of irrelevant structures or off-targets, which may induce toxicity (Salvador-Reyes and Luesch, 2015). Herbal formulation with two or more herbs in particular ratios have been able to exert a protective effect against numerous diseases like disorientation, forgetfulness, insomnia, unconsciousness, cramps, and seizures (Wu et al., 2011).

Ubiquitin-proteasome system and mechanism of action

SCA pathology involves unusual extension of cytosine-adenine-guanine (CAG) repeats resulting in the extended translation of the polyQ tract. Once the expansion overtakes a particular threshold, it induces the anomalous configuration of the encoded polyQ protein resulting in protein accumulations and nuclear inclusions, a pathological feature of SCA (McIntosh et al., 2017). The cellular existence of polyQ aggregates provokes cell death (Williams and Paulson, 2008; Dantuma and Bott, 2014). Interestingly, the ubiquitin-proteasome system (UPS) regulates intracellular protein homeostasis by eliminating misfolded and imperfect proteins (Dantuma and Bott, 2014). In addition, it also regulates cell cycle progression and apoptosis (Bader and Steller, 2009). Several lines of evidence have shown that polyQ proteins are vulnerable to ubiquitination and proteasome degradation

(Cummings et al., 1999; Matilla et al., 2001; Chai et al., 2004; Duenas et al., 2006). Nevertheless, growth of polyQ aggregates substantially hampered the activity of UPS in SCA (Chen et al., 2019a,b; Chou et al., 2015) by deubiquitylation resulting in increased growth of aggregates (Lew et al., 2020). Hence, reestablishing UPS action by natural remedies probably acts as a potential therapeutic arsenal against polyQ-provoked SCA.

Mounting evidence has shown the antiaggregation efficacy of catalpol isolated from *Rehmannia glutinosa* and daidzein and puerarin isolated from *Pueraria lobata* in in vitro models of SCA. Proteasome inhibitor such as carbobenzoxy-L-leucyl-L-leucyl-L-leucinal (MG132) showed protective efficacy in human embryonic kidney 293 cells (HEK-293) and neuroblastoma SH-SY5Y cells harboring mutant ataxin-3 proteins (Chen et al., 2019a). Pretreatment with 0.1 μM catalpol, puerarin, or daidzein for 24 h did not show a toxic effect against the cells and reinstated the UPS efficacy in turn augmenting the proteasome activity in the cells. These compounds also attenuated polyQ aggregation instigated by mutant ataxin-3 protein.

Antioxidant activity

When reactive oxygen species (ROS) production surpasses the antioxidant efficacy of intracellular antioxidant enzymes, oxidative damage usually takes place (Pizzino et al., 2017). PolyQ aggregate accumulation has been found to deteriorate the antioxidants such as glutathione, superoxide dismutase, and catalase, and enhance ROS generation and oxidative injury (Kim et al., 2003; Xiao et al., 2014). Patients suffering from SCA were found to have lower levels of glutathione and other thiols, which enhanced their vulnerability to oxidative injury (de Assis et al., 2017; Pacheco et al., 2013). Consequently, ROS-provoked DNA damage and cell death may happen resulting in motor dysfunctions linked to the disorder (Giuliano et al., 2003; Bertoni et al., 2011; Chen et al., 2014). In addition, ROS-augmented protein aggregation further enhances ROS production (Chen et al., 2014). Genipin, geniposide, crocin, and the aqueous extract of *Gardenia jasminoides* exhibited antioxidant efficacy in in vitro models of SCA such as HEK-293 and SH-SY5Y cells harboring mutant ataxin-3 proteins (Chang et al., 2014). These studies are in accordance with the findings of Chen et al. (2014) that showed a similar antioxidant activity of aqueous extract of *Glycyrrhiza inflata* and compounds like licochalcone A and ammonium glycyrrhizinatein SCA in in vitro models of HEK-293 and SH-SY5Y cells harboring mutant ataxin-3 proteins.

Pohl et al. (2019) examined the antioxidant efficacy of ethanolic extract of rapeseed pomace obtained during Soxhlet extraction of *Brassica napus*. In vivo administration of ethanolic extract (1–5 mg/mL) in the SCA model of transgenic *Caenorhabditis elegans* with mutant ataxin-3 protein exhibited no toxicity identified by food clearance assay. This treatment initiated the expression of glutathione *S*-transferase 4 and superoxide dismutase 3, which are entailed in glutathione metabolism and mitochondrial manganese SOD synthesis. Moreover, ethanolic extract ameliorated motor function in transgenic *C. elegans* compared to wild type as noticed by a motility assay.

Shaoyao Gancao Tang, a Chinese traditional herbal medicine, comprises *Paeonia lactiflora* and *Glycyrrhiza uralensis* in a 1:1 ratio. Chen et al. (2019c) evaluated the antioxidant activities of the formulation in in vitro models of SCA using HEK 293 and SHSY5Y cells expressing mutant TATA-box binding protein and in vivo transgenic mouse model of SCA. Zhengan Xifeng Tang, another Chinese traditional herbal medicine, comprises multifarious mixtures of medicinal plants. Okabe et al. (2007) investigated the protective efficacy of the formulation in a 60-year-old female suffering from typical SCA. Oral intake of the medicine for 2 months mitigated ataxia symptoms, such as vertigo and titubation, and ameliorated motor functions. However, 1 year after treatment was stopped, the ataxia symptoms reappeared despite the decline in symptoms after readministration of the formulation. Accordingly, results of Okabe (2018) in a 71-year-old female suffering from SCA coincided with earlier studies. Despite the lack of information on the mechanism of the formulation, antioxidant and anticonvulsant activity were hypothesized to play a key role in protective efficacy (Okabe et al., 2007; Okabe, 2018).

Ginkgo biloba

The *Ginkgo biloba* extract, EGb 761, has flavonoids and terpenoids, and substantially combats NDs such as Alzheimer's disease (AD) and Parkinson's disease (Ramassamy, 2006; Huang et al., 2016). EGb 761 treatment attenuated excitotoxicity and calcium entry, and mitigated the expression of apoptotic markers in glutamate-treated SH-SY5Y neuroblastoma cells. In addition, intraperitoneal injection of 100 mg/kg ameliorated motor deficits of SCA 17 transgenic mice (Huang et al., 2016). A wealth of studies has shown that *Ginkgo biloba* extract, EGb 761 with antioxidant activity, protects against ischemia, oxidative stress, and β-amyloid toxicity. It is extensively used in NDs such as AD, Parkinson's disease, and dementia (Oyama et al., 1996;

Bastianetto et al., 2000; Diamond et al., 2000; Chandrasekaran et al., 2001; Huang et al., 2016). Growing lines of evidence have shown that quality control is probably not sufficient for polyQ proteins, and organizing accurate folding of these proteins and averting their aggregation can be a potential therapeutic avenue (Nakamura et al., 2001; Ross and Poirier, 2004; Park et al., 2013; Huang et al., 2016).

Bacopa monnieri

Bacopa monnieri is a robust Indian medicinal herb generally called "Brahmi," belonging to the family Scrophulariaceae. It is commonly used to treat manifold diseases. This annual herb has ample bioactive molecules and has acted as a memory booster for 3000 years (Anonymous 2001; Gohil and Patel, 2010). It is also used as an alternative and complementary medicine in several countries. Nevertheless, there is scant research on the molecular mechanism of NDs treatment (Jeyasri et al., 2020). The Bacopa herb, usually known as a nootropic herb, facilitates the mending of damaged neurons, neuronal synthesis, reestablishment of synaptic activity, and ameliorates brain function. It possesses alkaloid brahmine, nicotine, herpestine, bacosides A and B, saponins A, B, and Cbetulinic acid, D-mannitol, triterpenoid saponins, stigmastanol, stigmasterol, aspartic acid, glutamic acid, and serine and pseudojujubogenin glycoside, β-sitosterol, and α-alanine (Devishree et al., 2017). Multifarious mechanisms of cognitive protection hitherto discovered are acetylcholinesterase (AChE) inhibition, antioxidant neuroprotection, neurotransmitter modulation (acetylcholine (ACh)), 5-hydroxytryptamine (5-HT), dopamine, choline acetyltranserase activation, and enhanced cerebral blood circulation (Aguiar and Borowski, 2013). Although not investigated until now, it has been hypothesized that associating *Bacopa monnieri* with a stimulant combats malaise (Ramasamy et al., 2015). Using chemininformatics, it has been shown that the 52 compounds of *Bacopa monnieri* can substantially target 780 human targets. Accordingly, novel molecules for AD and SCA treatment by suppressing AChE, β-amyloid accumulation in the brain, and degenerative changes in the cerebellum and spinal cord, respectively, have been found.

Chionanthus retusus Lindl. et Paxton (C. retusus)

Chionanthus retusus Lindl. et Paxton (*C. retusus*) encompasses a plethora of deciduous trees spread across tropical and subtropical countries such as China,

Japan, North Korea, and South Korea (Arias et al., 2011; Pei et al., 2020). The gentle leaves have been extensively used in medicinal tea or beverages due to the availability of plenty of polyphenols, flavonoids, and functional polysaccharides (Lee et al., 2019). Hot water (90–100°C) extraction profoundly liberates the water-soluble components (tea polyphenols, theophylline, amino acids, and vitamins) from roasted leaves while primary polysaccharides (cellulose and hemicellulose) and lignin are retained in the residual solids (Liu et al., 2018; Fu et al., 2020). The overproduction of ROS in cells induces oxidative stress, which is an underlying cause of diseases such as cancer, diabetes, inflammatory diseases, and NDs (Waris and Ahsan, 2006). The unusual genetic expansion of polyQ tract leads to NDs, such as Huntington's disease, and many subtypes of SCA (Lieberman et al., 2019). PolyQ protein accumulation profoundly perturbs regular biological functions and develops polyQ neurodegeneration (Adegbuyiro et al., 2017). Lignin-carbohydrate complexes independent carbohydrate (LCC-C-CR), lignin enriched (LCC-L-CR) LCC preparations, and the carbohydrate-free lignin preparation (Lignin-CR) isolated from the residual solids of *C. retuses* showed radical scavenging efficacy and polyQ protein aggregation inhibitory activity in in vitro mammalian cells and in vivo zebra fish models (Pei et al., 2020).

Azadiradione

Azadiradione, an active component of *Azadirachta indica*, has been found to be an inducer of HSF1 activity. *A. indica*, also called Neem, exhibits antiinflammatory and antianxiety activity, and augmentation of cognitive ability (Alam et al., 2012; Raghavendra et al., 2013; Nelson et al., 2016). Studies have shown that Azadiradione substantially improves polyQ protein-provoked toxicity in cellular and fruit fly models. The protective effects were associated with Azadiradione-prompted upregulation of heat shock protein chaperone (HSP) expression regulated by direct interaction with the HSF1 protein. Azadiradione connects with the purified human heat shock factor 1 with greater specificity. These findings support that Azadiradione is a promising molecule that can substantially combat NDs, which are prevalent globally (Nelson et al., 2016).

The primary mediator of heat shock response (HSR), heat shock factor 1 (HSF1) in normal cells, is located in the cytoplasm in a suppressive complex associated with HSP90, p23, immunophilin, HSP70, and HSP40 (Ali et al., 1998; Shi et al., 1998; Zou et al., 1998; Bharadwaj et al., 1999).

Enhanced temperature or proteotoxic stress leads to the dissociation of this complex and facilitates the accumulation of monomeric HSF1 molecules into DNA binding-capable homotrimer form. The role of ribonucleoprotein complexes in the trimerization and activation of HSF1 has been unraveled (Shamovsky and Nudler, 2006; Vera et al., 2014). Chemical attenuation of HSP90 and proteasome also leads to HSF1 activation (Zou et al., 1998; McDonough and Patterson, 2003).

Mounting evidence has shown the neuroprotective activity of Azadiradione in a fly model of polyQ diseases by HSF activation, leading to increased expression of HSP70 at transcript and protein levels (Kazemi-Esfarjani and Benzer, 2000; Arya et al., 2007; Mallik and Lakhotia, 2010; Nelson et al., 2016). The antioxidant activity was observed to decrease protein accumulation in fly and mouse models, which prompted its use to combat polyQ diseases (Ehrnhoefer et al., 2006; Melkani et al., 2013).

Conclusions and future perspectives

SCA pathology involves unusual extension of CAG repeats resulting in the extended translation of the polyQ tract. Despite extensive research on NDs and their therapeutics, there is no substantial treatment since current treatment offers moderate relief of the symptoms (Cookson, 2010; Neef et al., 2011; Nelson et al., 2016). The neuroprotective activity of Azadiradione in a fly model of polyQ diseases by HSF activation leads to increased expression of HSP70 at transcript and protein levels. Extensive research studies are to be carried out to elucidate the molecular mechanisms of therapeutic efficacy of these plant-based molecules. There is an urgent need to optimize the structures of biological compounds to enhance their affinity to disease targets and wane activity toward off-targets (Phang et al., 2021). The drug efficacy of herbal medicine may be because of individual and synergistic activity of phytomolecules. It is indeed a biological riddle to study such molecular mechanisms.

References

Adegbuyiro, A., Sedighi, F., Pilkington IV, A.W., Groover, S., Legleiter, J., 2017. Proteins containing expanded polyglutamine tracts and neurodegenerative disease. Biochemistry 56, 1199–1217.

Aguiar, S., Borowski, T., 2013. Neuropharmacological review of the nootropic herb Bacopa monnieri. Rejuvenation Res. 16, 313–326.

Akbar, U., Ashizawa, T., 2015. Ataxia. Neurol. Clin. 33, 225–248.

Alam, A., Haldar, S., Thulasiram, H.V., Kumar, R., Goyal, M., Iqbal, M.S., et al., 2012. Novel anti-inflammatory activity of epoxyazadiradione against macrophage migration inhibitory factor: inhibition of tautomerase and proinflammatory activities of macrophage migration inhibitory factor. J. Biol. Chem. 287, 24844–24861.

Albernaz, P.L.M., Zuma e Maia, F.Z., Carmona, S., Cal, R.V.R., Zalazar, G., 2019. Ataxia. In: The New Neurotology: A Comprehensive Clinical Guide. Springer, Cham, pp. 181–191.

Ali, A., Bharadwaj, S., O'Carroll, R., Ovsenek, N., 1998. HSP90 interacts with and regulates the activity of heat shock factor 1 in Xenopus oocytes. Mol. Cell. Biol. 18, 4949–4960.

Anonymous, 2001. The Ayurvedic Pharmacopoeia of India. Government of India, New Delhi, India. Ministry of Health and Family Welfare: New Delhi, India, Part-I, vol. III.

Arias, R.S., Techen, N., Rinehart, T.A., Olsen, R.T., Kirkbride, J.H., Scheffler, B.E., 2011. Development of simple sequence repeat markers for Chionanthus retusus (Oleaceae) and effective discrimination of closely related taxa. Hortic. Sci. 46, 23–29.

Arya, R., Mallik, M., Lakhotia, S.C., 2007. Heat shock genes – integrating cell survival and death. J. Biosci. 32, 595–610.

Auluck, P.K., Bonini, N.M., 2002. Pharmacological prevention of Parkinson disease in Drosophila. Nat. Med. 8, 1185–1186.

Ayaz, M., Ullah, F., Sadiq, A., Kim, M.O., Ali, T., 2019. Natural products-based drugs: potential therapeutics against Alzheimer's disease and other neurological disorders. Front. Pharmacol. 10, 1417.

Bader, M., Steller, H., 2009. Regulation of cell death by the ubiquitin–proteasome system. Curr. Opin. Cell Biol. 21, 878–884.

Bastianetto, S., Ramassamy, C., Doré, S., Christen, Y., Poirier, J., Quirion, R., 2000. The *Ginkgo biloba* extract (EGb 761) protects hippocampal neurons against cell death induced by beta-amyloid. Eur. J. Neurosci. 12, 1882–1890.

Bates, G., 2003. Huntingtin aggregation and toxicity in Huntington's disease. Lancet 361, 1642–1644.

Bertoni, A., Giuliano, P., Galgani, M., Rotoli, D., Ulianich, L., Adornetto, A., et al., 2011. Early and late events induced by polyQ-expanded proteins: identification of a common pathogenic property of polyQ-expanded proteins. J. Biol. Chem. 286, 4727–4741.

Bharadwaj, S., Ali, A., Ovsenek, N., 1999. Multiple components of the HSP90 chaperone complex function in regulation of heat shock factor 1 *in vivo*. Mol. Cell. Biol. 19, 8033–8041.

Chai, Y., Berke, S.S., Cohen, R.E., Paulson, H.L., 2004. Poly-ubiquitin binding by the polyglutamine disease protein ataxin-3 links its normal function to protein surveillance pathways. J. Biol. Chem. 279, 3605–3611.

Chandrasekaran, K., Mehrabian, Z., Spinnewyn, B., Drieu, K., Fiskum, G., 2001. Neuroprotective effects of bilobalide, a component of the *Ginkgo biloba* extract (EGb 761), in gerbil global brain ischemia. Brain Res. 922, 282–292.

Chang, K.-H., Chen, W.-L., Wu, Y.-R., Lin, T.-H., Wu, Y.-C., Chao, C.-Y., et al., 2014. Aqueous extract of Gardenia jasminoides targeting oxidative stress to reduce polyQ aggregation in cell models of spinocerebellar ataxia 3. Neuropharmacology 81, 166–175.

Chen, C.-M., Weng, Y.-T., Chen, W.-L., Lin, T.-H., Chao, C.-Y., Lin, C.-H., et al., 2014. Aqueous extract of Glycyrrhiza inflata inhibits aggregation by upregulating PPARGC1A and NFE2L2–ARE pathways in cell models of spinocerebellar ataxia 3. Free Radic. Biol. Med. 71, 339–350.

Chen, I.C., Chang, C.N., Chen, W.L., Lin, T.H., Chao, C.Y., Lin, C.H., et al., 2019a. Targeting ubiquitin proteasome pathway with traditional Chinese medicine for treatment of spinocerebellar ataxia type 3. Am. J. Chin. Med. 47, 63–95.

Chen, I.C., Chang, K.H., Chen, Y.J., Chen, Y.C., Lee-Chen, G.J., Chen, C.M., et al., 2019b. *Pueraria lobata* and daidzein reduce cytotoxicity by enhancing ubiquitin proteasome system function in SCA3-iPSC-derived neurons. Oxidative Med. Cell. Longev. 2019, 8130481.

Chen, C.M., Chen, W.L., Hung, C.T., Lin, T.H., Lee, M.C., Chen, I.C., et al., 2019c. Shaoyao Gancao Tang (SG-Tang), a formulated Chinese medicine, reduces aggregation and exerts

neuroprotection in spinocerebellar ataxia type 17 (SCA17) cell and mouse models. Aging 11, 986–1007.
Chou, A.H., Chen, Y.L., Chiu, C.C., Yuan, S.J., Weng, Y.H., Yeh, T.H., et al., 2015. T1-11 and JMF1907 ameliorate polyglutamine expanded ataxin-3-induced neurodegeneration, transcriptional dysregulation and symptoms of ataxia in the SCA3 transgenic mouse. Neuropharmacology 99, 308–317.
Chung, C.G., Kwon, M.J., Jeon, K.H., Hyeon, D.Y., Han, M.H., Park, J.H., et al., 2017. Golgi outpost synthesis impaired by toxic polyglutamine proteins contributes to dendritic pathology in neurons. Cell Rep. 20, 356–369.
Cookson, M.R., 2010. The role of leucine-rich repeat kinase 2 (LRRK2) in Parkinson's disease. Nat. Rev. Neurosci. 11, 791–797.
Cummings, C.J., Reinstein, E., Sun, Y., Antalffy, B., Jiang, Y.H., Ciechanover, A., et al., 1999. Mutation of the E6-AP ubiquitin ligase reduces nuclear inclusion frequency while accelerating polyglutamine induced pathology in SCA1 mice. Neuron 24, 879–892.
Dantuma, N.P., Bott, L.C., 2014. The ubiquitin-proteasome system in neurodegenerative diseases: precipitating factor, yet part of the solution. Front. Mol. Neurosci. 7, 70.
Davidson-Hunt, I., 2000. Ecological ethnobotany: stumbling toward new practices and paradigms. MASA J. 16, 1–13.
de Assis, A.M., Saute, J.A.M., Longoni, A., Haas, C.B., Torrez, V.R., Brochier, A.W., et al., 2017. Peripheral oxidative stress biomarkers in spinocerebellar ataxia type 3/Machado–Joseph disease. Front. Neurol. 8, 485.
Dendi, V.S., Obulesu, M., 2020. Neuroprotective role of resveratrol in Alzheimer's disease. In: Bhattacharjee, A., Ramakrishna, A., Obulesu, M. (Eds.), Phytomedicine and Alzheimer's Disease. CRC Press.
Devishree, R.A., Kumar, S., Jain, A.R., 2017. Short term effect of Bacopa monnieri on memory—a brief review. J. Pharm. Res. 11, 1447–1450.
Diamond, B.J., Shiflett, S.C., Feiwel, N., Mtheis, R.J., Noskin, O., Richards, J.A., et al., 2000. *Ginkgo biloba* extract: mechanisms and clinical indications. Arch. Phys. Med. Rehabil. 81, 668–678.
Duenas, A.M., Goold, R., Giunti, P., 2006. Molecular pathogenesis of spinocerebellar ataxias. Brain 129, 1357–1370.
Ehrnhoefer, D.E., Duennwald, M., Markovic, P., Wacker, J.L., Engemann, S., Roark, M., et al., 2006. Green tea (−)-epigallocatechin-gallate modulates early events in huntingtin misfolding and reduces toxicity in Huntington's disease models. Hum. Mol. Genet. 15, 2743–2751.
Fu, Y., Li, F., Ding, Y., Li, H.Y., Xiang, X.R., Ye, Q., et al., 2020. Polysaccharides from loquat (Eriobotrya japonica) leaves: impacts of extraction methods on their physicochemical characteristics and biological activities. Int. J. Biol. Macromol. 146, 508–517.
Fujigasaki, H., Martin, J.J., DeDeyn, P.P., Camuzat, A., Deffond, D., Stevanin, G., et al., 2001. CAG repeat expansion in the TATA box-binding protein gene causes autosomal dominant cerebellar ataxia. Brain 124, 1939–1947.
Giuliano, P., de Cristofaro, T., Affaitati, A., Pizzulo, G.M., Feliciello, A., Criscuolo, C., et al., 2003. DNA damage induced by polyglutamine-expanded proteins. Hum. Mol. Genet. 12, 2301–2309.
Gohil, K.J., Patel, J.A., 2010. A review on Bacopa monniera: current research and future prospects. Int. J. Green Pharm. 4, 1–9.
Graveland, G.A., Williams, R.S., DiFiglia, M., 1985. Evidence for degenerative and regenerative changes in neostriatal spiny neurons in Huntington's disease. Science 227, 770–773.
Groth, C.L., Berman, B.D., 2018. Spinocerebellar ataxia 27: a review and characterization of an evolving phenotype. Tremor Other Hyperkinet. Mov. 8, 534.
Harvey, A.L., Edrada-Ebel, R., Quinn, R.J., 2015. The re-emergence of natural products for drug discovery in the genomics era. Nat. Rev. Drug Discov. 14, 111–129.

Huang, D.S., Lin, H.Y., Lee-Chen, G.J., Hsieh-Li, H.M., Wu, C.H., Lin, J.Y., 2016. Treatment with a *Ginkgo biloba* extract, EGb 761, inhibits excitotoxicity in an animal model of spinocerebellar ataxia type 17. Drug Des. Devel. Ther. 10, 723–731.

Hubener, J., Weber, J.J., Richter, C., Honold, L., Weiss, A., Murad, F., et al., 2013. Calpain-mediated ataxin-3 cleavage in the molecular pathogenesis of spinocerebellar ataxia type 3 (SCA3). Hum. Mol. Genet. 22, 508–518.

Jeyasri, R., Muthuramalingam, P., Suba, V., Ramesh, M., Chen, J.T., 2020. *Bacopa monnieri* and their bioactive compounds inferred multi-target treatment strategy for neurological diseases: a cheminformatics and system pharmacology approach. Biomol. Ther. 10, 536.

Kazemi-Esfarjani, P., Benzer, S., 2000. Genetic suppression of polyglutamine toxicity in Drosophila. Science 287, 1837–1840.

Khaleel, M., Obulesu, M., 2020. Neuroprotective role of resveratrol in Alzheimer's disease. In: Bhattacharjee, A., Ramakrishna, A., Obulesu, M. (Eds.), Phytomedicine and Alzheimer's Disease. CRC Press.

Khan, A., Jahan, S., Imtiyaz, A., Alshahrani, S., Antar Makeen, H., Alshehri, BM., et al., 2020. Neuroprotection: targeting multiple pathways by naturally occurring phytochemicals. Biomedicines 8, 284.

Kim, S.J., Kim, T.S., Hong, S., Rhim, H., Kim, I.Y., Kang, S., 2003. Oxidative stimuli affect polyglutamine aggregation and cell death in human mutantataxin-1-expressing cells. Neurosci. Lett. 348, 21–24.

Klockgether, T., Mariotti, C., Paulson, H.L., 2019. Spinocerebellar Ataxia. Nat. Rev. Dis. Primers 5, 24.

Koch, P., Breuer, P., Peitz, M., Jungverdorben, J., Kesavan, J., Poppe, D., et al., 2011. Excitation-induced ataxin-3 aggregation in neurons from patients with Machado-Joseph disease. Nature 480, 543–546.

Koeberle, A., Werz, O., 2014. Multi-target approach for natural products in inflammation. Drug Discov. Today 19, 1871–1882.

Kulkarni, V.A., Firestein, B.L., 2012. The dendritic tree and brain disorders. Mol. Cell. Neurosci. 50, 10–20.

Kwon, M.J., Han, M.H., Bagley, J.A., Hyeon, D.Y., Ko, B.S., Lee, Y.M., et al., 2018. Coiled-coil structure-dependent interactions between polyQ proteins and Foxo lead to dendrite pathology and behavioral defects. Proc. Natl. Acad. Sci. U. S. A. 115, E10748–E10757.

Labbadia, J., Morimoto, R.I., 2015. The biology of proteostasis in aging and disease. Annu. Rev. Biochem. 84, 435–464.

Lee, S.B., Bagley, J.A., Lee, H.Y., Jan, L.Y., Jan, Y.N., 2011. Pathogenic polyglutamine proteins cause dendrite defects associated with specific actin cytoskeletal alterations in Drosophila. Proc. Natl. Acad. Sci. U. S. A. 108, 16795–16800.

Lee, Y.G., Lee, H., Jung, J.W., Seo, K.H., Lee, D.Y., Kim, H.G., et al., 2019. Flavonoids from *Chionanthus retusus* (Oleaceae) flowers and their protective effects against glutamate-induced cell toxicity in HT22 cells. Int. J. Mol. Sci. 20, 3517.

Lew, S.Y., Yow, Y.Y., Lim, L.W., Wong, K.-H., 2020. Antioxidant-mediated protective role of *Hericium erinaceus* (Bull.: Fr.) Pers. against oxidative damage in fibroblasts from Friedreich's ataxia patient. Food Sci. Technol. 40 (Suppl. 1), 264–272.

Lieberman, A.P., Shakkottai, V.G., Albin, R.L., 2019. Polyglutamine repeats in neurodegenerative diseases. Annu. Rev. Pathol. 14, 1–27.

Liu, Y., Fu, Y., Shao, Z., Zhang, F., Qin, M., 2018. Effects of pH control with acetic acid/sodium acetate buffer on the hot-water extraction of poplar wood. J. Bioresour. Bioprod. 3, 9–13.

Maheshwari, M., Bhutani, S., Das, A., Mukherjee, R., Sharma, A., Kino, Y., et al., 2014. Dexamethasone induces heat shock response and slows down disease progression in mouse and fly models of Huntington's disease. Hum. Mol. Genet. 23, 2737–2751.

Mallik, M., Lakhotia, S.C., 2010. Modifiers and mechanisms of multi-system polyglutamine neurodegenerative disorders: lessons from fly models. J. Genet. 89, 497–526.

Matilla, A., Gorbea, C., Einum, D.D., Townsend, J., Michalik, A., van Broeckhoven, C., et al., 2001. Association of ataxin-7 with the proteasome subunit S4 of the 19S regulatory complex. Hum. Mol. Genet. 10, 2821–2831.

McDonough, H., Patterson, C., 2003. CHIP: a link between the chaperone and proteasome systems. Cell Stress Chaperones 8, 303–308.

McIntosh, C.S., Aung-Htut, M.T., Fletcher, S., Wilton, S.D., 2017. Polyglutamine ataxias: from clinical and molecular features to current therapeutic strategies. J. Genet. Syndr. Gene. Ther. 8, 1–17.

Melkani, G.C., Trujillo, A.S., Ramos, R., Bodmer, R., Bernstein, S.I., Ocorr, K., 2013. Huntington's disease induced cardiac amyloidosis is reversed by modulating protein folding and oxidative stress pathways in the Drosophila heart. PLoS Genet. 9, e1004024.

Morley, J.F., Brignull, H.R., Weyers, J.J., Morimoto, R.I., 2002. The threshold for polyglutamine expansion protein aggregation and cellular toxicity is dynamic and influenced by aging in Caenorhabditis elegans. Proc. Natl. Acad. Sci. U. S. A. 99, 10417–10422.

Nakamura, K., Jeong, S.Y., Uchihara, T., Anno, M., Nagashima, K., Nagashima, T., et al., 2001. SCA 17, a novel autosomal dominant cerebellar ataxia caused by an expanded polyglutamine in TATA-binding protein. Hum. Mol. Genet. 10, 1441–1448.

Neef, D.W., Turski, M.L., Thiele, D.J., 2010. Modulation of heat shock transcription factor 1 as a therapeutic target for small molecule intervention in neurodegenerative disease. PLoS Biol. 8, e1000291.

Neef, D.W., Jaeger, A.M., Thiele, D.J., 2011. Heat shock transcription factor 1 as a therapeutic target in neurodegenerative diseases. Nat. Rev. Drug Discov. 10, 930–944.

Nelson, V.K., Ali, A., Dutta, N., Ghosh, S., Jana, M., Ganguli, A., et al., 2016. Azadiradione ameliorates polyglutamine expansion disease in *Drosophila* by potentiating DNA binding activity of heat shock factor 1. Oncotarget 7, 78281–78296.

Obulesu, M., 2020. Quercetin in Alzheimer's disease. In: Bhattacharjee, A., Ramakrishna, A., Obulesu, M. (Eds.), Phytomedicine and Alzheimer's Disease. CRC Press.

Obulesu, M., 2021. Turmeric and Curcumin in the Treatment of Neurodegenerative Diseases. Elsevier.

Okabe, T., 2018. Remission induction of spinocerebellar ataxia 6 with medicinal herbs. J. Aging Neuropsychol. 108, 1–3.

Okabe, T., Fujisawa, M., Sekiya, T., Ichikawa, Y., Goto, J., 2007. Successful treatment of spinocerebellar ataxia 6 with medicinal herbs. Geriatr. Gerontol. Int. 7, 195–197.

Oyama, Y., Chikahisa, L., Ueha, T., Kanemaru, K., Noda, K., 1996. *Ginkgo biloba* extract protects brain neurons against oxidative stress induced by hydrogen peroxide. Brain Res. 712, 349–352.

Pacheco, L.S., da Silveira, A.F., Trott, A., Houenou, L.J., Algarve, T.D., Bello, C., et al., 2013. Association between Machado-Joseph disease and oxidative stress biomarkers. Mutat. Res. 757, 99–103.

Park, S.H., Kukushkin, Y., Gupta, R., Chen, T., Konagai, A., Hipp, M.S., et al., 2013. PolyQ proteins interfere with nuclear degradation of cytosolic proteins by sequestering the Sis1p chaperone. Cell 154, 134–145.

Pei, W., Chen, Z.S., Chan, H.Y.E., Zheng, L., Liang, C., Huang, C., 2020. Isolation and identification of a novel anti-protein aggregation activity of lignin-carbohydrate complex from Chionanthus retusus leaves. Front. Bioeng. Biotechnol. 8, 573991.

Phang, M.W.L., Lew, S.Y., Chung, I., Lim, W.K., Lim, L.W., Wong, K.H., 2021. Therapeutic roles of natural remedies in combating hereditary ataxia: a systematic review. Chin. Med. 16, 15.

Pizzino, G., Irrera, N., Cucinotta, M., Pallio, G., Mannino, F., Arcoraci, V., et al., 2017. Oxidative stress: harms and benefits for human health. Oxidative Med. Cell. Longev. 2017, 8416763.

Pohl, F., Kong Thoo Lin, P., 2018. The potential use of plant natural products and plant extracts with antioxidant properties for the prevention/treatment of neurodegenerative diseases: in vitro, in vivo and clinical trials. Molecules 23, 3283.

Pohl, F., Teixeira-Castro, A., Costa, M.D., Lindsay, V., Fiúza-Fernandes, J., Goua, M., et al., 2019. GST-4-dependent suppression of neurodegeneration in *C. elegans* models of Parkinson's and Machado-Joseph disease by rapeseed pomace extract supplementation. Front. Neurosci. 13, 1091.

Raghavendra, M., Maiti, R., Kumar, S., Acharya, S., 2013. Role of aqueous extract of Azadirachta indica leaves in an experimental model of Alzheimer's disease in rats. Int. J. Appl. Basic Med. Res. 3, 37–47.

Ramasamy, S., Chin, S.P., Sukumaran, S.D., Buckle, M.J.C., Kiew, L.V., Chung, L.Y., 2015. In silico and in vitro analysis of bacoside a aglycones and its derivatives as the constituents responsible for the cognitive effects of Bacopa monnieri. PLoS One 10, e0126565.

Ramassamy, C., 2006. Emerging role of polyphenolic compounds in the treatment of neurodegenerative diseases: a review of their intracellular targets. Eur. J. Pharmacol. 545, 51–64.

Ross, C., Poirier, M., 2004. Protein aggregation and neurodegenerative disease. Nat. Med. 10, 10–17.

Salvador-Reyes, L.A., Luesch, H., 2015. Biological targets and mechanisms of action of natural products from marine cyanobacteria. Nat. Prod. Rep. 32, 478–503.

Sarva, H., Shanker, V.L., 2014. Treatment options in degenerative cerebellar ataxia: a systematic review. Mov. Disord. Clin. Pract. 1, 291–298.

Schols, L., Bauer, P., Schmidt, T., Schulte, T., Riess, O., 2004. Autosomal dominant cerebellar ataxias: clinical features, genetics, and pathogenesis. Lancet Neurol. 3, 291–304.

Shamovsky, I., Nudler, E., 2006. Gene control by large noncoding RNAs. Sci. STKE 2006, 40.

Shi, Y., Mosser, D.D., Morimoto, R.I., 1998. Molecular chaperones as HSF1-specific transcriptional repressors. Genes Dev. 12, 654–666.

Simoes, A.T., Goncalves, N., Koeppen, A., Deglon, N., Kugler, S., Duarte, C.B., et al., 2012. Calpastatin-mediated inhibition of calpains in the mouse brain prevents mutant ataxin 3 proteolysis, nuclear localization and aggregation, relieving Machado-Joseph disease. Brain 135, 2428–2439.

Sotrel, A., Williams, R.S., Kaufmann, W.E., Myers, R.H., 1993. Evidence for neuronal degeneration and dendritic plasticity in cortical pyramidal neurons of Huntington's disease: a quantitative Golgi study. Neurology 43, 2088–2096.

Stuart, G.J., Spruston, N., 2015. Dendritic integration: 60 years of progress. Nat. Neurosci. 18, 1713–1721.

Sullivan, R., Yau, W.Y., O'Connor, E., Houlden, H., 2018. Spinocerebellar ataxia: an update. J. Neurol. 266, 533–544.

Teive, H.A.G., Ashizawa, T., 2015. Primary and secondary ataxias. Curr. Opin. Neurol. 28, 413–422.

Vera, M., Pani, B., Griffiths, L.A., Muchardt, C., Abbott, C.M., Singer, R.H., et al., 2014. The translation elongation factor eEF1A1 couples transcription to translation during heat shock response. Elife 3, e03164.

Waris, G., Ahsan, H., 2006. Reactive oxygen species: role in the development of cancer and various chronic conditions. J. Carcinog. 5, 14.

Williams, A.J., Paulson, H.L., 2008. Polyglutamine neurodegeneration: protein misfolding revisited. Trends Neurosci. 31, 521–528.

Wu, T.Y., Chen, C.P., Jinn, T.R., 2011. Traditional Chinese medicines and Alzheimer's disease. Taiwan. J. Obstet. Gynecol. 50, 131–135.

Xiao, L., Li, H., Zhang, J., Yang, F., Huang, A., Deng, J., et al., 2014. Salidroside protects *Caenorhabditis elegans* neurons from polyglutamine-mediated toxicity by reducing oxidative stress. Molecules 19, 7757–7769.

Zou, J., Guo, Y., Guettouche, T., Smith, D.F., Voellmy, R., 1998. Repression of heat shock transcription factor HSF1 activation by HSP90 (HSP90 complex) that forms a stress-sensitive complex with HSF1. Cell 94, 471–480.

Zuhlke, C., Hellenbroich, Y., Dalski, A., Kononowa, N., Hagenah, J., Vieregge, P., et al., 2001. Different types of repeat expansion in the TATA-binding protein gene are associated with a new form of inherited ataxia. Eur. J. Hum. Genet. 9, 160–164.

CHAPTER 7

Multifarious plant compounds and their protective efficacy in the treatment of spinal muscular atrophy

Introduction

Pathology

Spinal muscular atrophy (SMA) is a genetically heterogeneous class of atypical neuromuscular diseases (NDs) that has been a major cause of death in children until recently (Deutsch et al., 2021). It is the second most common genetic basis of infant mortality next to cystic fibrosis (Lunn and Wang, 2008; Mazzarella et al., 2019). The 5q SMA is an unusual ND that results in constant atrophy and deterioration of skeletal muscles because of the degeneration of motor neurons (MNs) (Melki, 2017). It is categorized by loss of the MNs in the anterior horn of the spinal cord, leading to atrophy and fragility of the voluntary muscles of the limbs and trunk. It has been the major genetic cause of child deaths with an occurrence of around one in 11,000 births (Sugarman et al., 2012). The gene responsible for survival motor neuron (SMN) was discovered in 1990 by two different groups (Gilliam et al., 1990; Melki et al., 1990) and was found to be localized on chromosome 5q13; in 1996, it was found as a disease-inducing gene. The SMN protein is generated by two genes, the telomeric SMN1 gene, which is removed or broken in patients with SMA, and the centromeric SMN2 gene that varies from SMN1 by five nucleotides and exists in multiple copies. SMN1 generates full-length transcripts, whereas in 90% of patients SMN2 generates transcripts without exon 7 because of the C-to-T mutation that generates an exon-splicing suppressor sequence (Lefebvre et al., 1995; Lorson and Androphy, 2000). SMA also disturbs skeletal muscle, kidney, heart, pancreas, liver, bone, spleen, connective tissues, and immune systems (Yeo and Darras, 2020). The pathology is linked to a deficit in the survival motor neuron 1 (Smn1) that is universally expressed within and outside the nervous system

Plant Extracts in Neurodegenerative Diseases
https://doi.org/10.1016/B978-0-323-95762-5.00007-2

(Lefebvre et al., 1995; Young et al., 2000). Although a decrease in SMN is observed in all tissues, the characteristic of SMA is the specific loss of spinal motor neurons (MNs) in both humans and mouse models of the disease (Burghes and Beattie, 2009; Simon et al., 2017).

SMA is characterized into five types (0–IV) depending on the age of the patient and the intensity of the disease (Fuller et al., 2016; Khan et al., 2020). Advent of innovative therapies in the treatment of SMA altered the natural course of the disease by altering the expression or substituting mutated genes implicated in the progression of SMA. These treatment options include nusinersen (Chiriboga, 2017) (an antisense oligonucleotide that alters mRNA splicing), onasemnogene, abeparvovec (Al-Zaidy et al., 2019) (gene replacement therapy), and risdiplam (Poirier et al., 2018) (a small molecule that amends mRNA splicing). Nusinersen is an antisense oligonucleotide that initiates the inclusion of exon 7 into mRNA transcripts of SMN2. It combines into an intronic splice site in intron 7 of SMN2 and attenuates the action of other splicing factors, thereby initiating exon 7 integration into the mRNA. This results in the generation of an entirely functional SMN protein. Antisense oligonucleotides fail to span the blood-brain barrier, therefore they must be introduced intrathecally (Singh et al., 2006; Hua et al., 2010; Rigo et al., 2012).

The underlying cause of most of SMA is degeneration of MN proteins due to the homozygous deletion or point mutation in the SMN1 gene. SMN2 also encodes SMN protein but generates a small part of functional full-length proteins. Compared to SMN1, SMN2 possesses a single, silent C-to-T transition in exon 7 that initiates exon 7 skipping through alternative splicing. Thus, the obtained truncated SMN protein is not active and swiftly undergoes degradation (Groen et al., 2018). The SMN2 copy number is usually contrariwise related to clinical severity (Kolb et al., 2017; Wadman et al., 2017).

Nusinersen therapy

Nusinersen intrathecal application was the earliest therapy for the treatment of SMA approved by the Food and Drug Administration (FDA) and the European Medicines Agency (EMA) (Corey, 2017; Ramdas and Servais, 2020). Therefore, in a single-centered study, urine, serum, and liquor samples were taken from 25 male and female pediatric patients with SMA before and after 2 months of nusinersen therapy and urine samples from an identical healthy cohort ($n = 125$). Metabolomes were investigated with targeted metabolomics utilizing 600 MHz ^{1}H NMR, parametric and nonparametric multivariate statistical analyses, and machine learning and modeling. Medical evaluation before and after nusinersen therapy demonstrated remarkable amelioration in movement, posture, and strength in line with multifarious medical tests. Noteworthy differences were not noticed in metabolomes before and after nusinersen therapy in urine, serum, and liquor samples using a collection of statistical and machine learning approaches. Compared to a healthy cohort, ^{1}H NMR metabolomes of SMA patients had a decreased number and concentration of urine metabolites and varied profoundly between males and females. Machine learning corroborated urinary creatinine as primarily important, differentiating SMA patients from the healthy cohort.

The nusinersen drug ameliorated motor function and enhanced the magnitude of muscle action potential of the ulnar and peroneal nerve. Autopsy studies demonstrated the presence of nusinersen in MNs all through the spinal cord and in the neurons of the brainstem (Finkel et al., 2016). Augmented motor function was noticeable as an enhanced capability to sit or walk alone (Chiriboga et al., 2016; De Vivo et al., 2019; Osredkar et al., 2021), enhanced bite strength (Kruse et al., 2020), and hand strength (De Wel et al., 2020). Despite the intricacy involved in the intrathecal introduction of nusinersen and the requirement of dexterity, it has been adequately tolerated and safe (Stolte et al., 2018). Nusinersen currently exists in several nations for several types of SMA patients based primarily on inclusion criteria and economy of the country. Nonetheless, there is a burgeoning need to corroborate nusinersen's capability for a thorough understanding of the changeability of effect and adverse effects in a larger cohort of SMA patients (Michelson et al., 2018).

Plant extracts

Plant products and their efficacy against SMA play an important role as SMA is a genetic disorder. *Brucea javanica* and its vital constituent bruceine D were observed to show SMN2 splicing-correcting property. This study demonstrated that they mended the splice deficiency and remarkably augmented the activity of SMN2. *B. javanica* could improve muscle defects (Baek et al., 2019; Khan et al., 2020). Triptolide is another bioactive compound that exhibited a noteworthy therapeutic efficacy against SMA in SMA-like mice. Triptolide extracted from *Tripterygium wilfordii* enhanced the generation of SMN protein and transcription of SMN2 in both cellular and animal models. Hsu et al. (2012) confirmed that triptolide is a regulator of SMN expression. Triptolide improved the expression of transcript and protein levels of live MNs in human SMA fibroblasts and ameliorated endurance in SMA-like mice.

Kiwi fruit

Kiwi fruit, a functional food, possesses abundant nutraceuticals. It has numerous health benefits of which its neuroprotective efficacy has been studied in rats with learning and memory dysfunction promoted by exposure to a variety of chemicals (Mazzarella et al., 2019). Green kiwi fruit extracts showed a potential protective efficacy in the *Caenorhabditis elegans* SMA model by averting the degeneration and death of MNs and the locomotion disability, while the gold species could not show such effects. The compounds exerting neuroprotection are molecules with a molecular weight more than 3 kDa found in the green but not in the yellow kiwi fruit. It has been used for generations in traditional Chinese medicine to treat deleterious diseases like cancer (Motohashi et al., 2002). Studies have explored its anticonstipation (Chang et al., 2010), cardiovascular disease protection (Recio-Rodriguez et al., 2015), antibacterial (Popovic et al., 2013), and antinematode (Stepek et al., 2005) activities. Kiwi fruit species *Actinidia arguta* exerted neuroprotective efficacy and ameliorated learning and memory dysfunction initiated by trimethyltin in a mouse model (Ha et al., 2015). Additionally, it has been shown that the Kiwi fruit *Actinidia chinensis*, cultivar Qinmei, substantially ameliorated learning and memory deficits initiated in rats by lead exposure (Xue et al., 2017).

The nematode *C. elegans* is the best model for the recognition and characterization of bioactive molecules (Parker et al., 2005; Canuelo et al., 2012; Ding et al., 2017; Shukla et al., 2012; Srivastava et al., 2017) since it facilitates the swift recognition of new disease-related genetic mutants (Illiano et al., 2017; Di Schiavi and Andrenacci, 2013). A novel *C. elegans* model to study SMA has been disclosed recently (Gallotta et al., 2016). The SMA model designed by Gallotta et al. (2016) shows an age-dependent deterioration of MNs identified as locomotory deficiency, the desertion of presynaptic and cytoplasmic fluorescent markers, and eventually the apoptosis of targeted neurons. In contrast to the other models of SMA (Di Giorgio et al., 2017; Singh et al., 2017), this new model facilitates neuron-specific silencing of *smn-1*, averting the systemic effects that avoid the estimation of the definite role of *smn-1* in neuron survival. Hence, this model system facilitates the overcoming of riddles because of the pleiotropic phenotypes and simultaneously explores the benefits of specific features of this nematode, like its small size, quick generation time, body transparency, and short lifespan, and aids in the recognition of the factors avoiding the death of the MNs in vivo (de Carlos Caceres et al., 2018). Based on the results of this study, the evaluation of high molecular weight compounds of green Kiwi fruit such as protease actinidin, kiwellin, thaumatin-like protein, and the nutraceutical peptide kissper with substantial antioxidant and antiinflammatory activity (Ciacci et al., 2014) has been done. Unfortunately, these molecules failed to show pronounced neuroprotective activity. However, the optimum concentrations of these proteins are likely to exert protective efficacy through synergistic effects of extracts. SOD has been recognized to be a potential macromolecule for the future evaluation of protective efficacy. A wealth of studies has shown multifarious SOD proteins with neuroprotective activity, like ischemia models (Huang et al., 2012; Jung et al., 2009; Kondo et al., 1997; Polazzi et al., 2013) and in *C. elegans* models of NDs (Liu et al., 2015; Luo et al., 2011). Despite the significant activity of SOD in Kiwi fruit (Giangrieco et al., 2016), the availability of purified protein is a major problem. Since the quantity of ingredients expressing biological activity differs in different varieties, neuroprotective efficacy exhibited by different green Kiwi fruit extracts will also be different. This is usually because of numerous factors influencing the concentration of Kiwi fruit constituents (Ciardiello et al., 2009; Giangrieco et al., 2016).

RNA-targeted therapy

Pre-mRNA splicing is a crucial process that generates functional mature RNA in eukaryotic cells. In this process, introns are excised, and exons are combined. Pre-mRNA splicing is promoted by a macromolecular ribonucleoprotein complex called the spliceosome. The primary spliceosome comprises five small nuclear RNAs (U1, U2, U4, U5, and U6) that associate with proteins to make small nuclear ribonucleoproteins (snRNP). These snRNPs identify conserved sequence elements at 5′ splice sites (5′ss), 3′ss, and branch points (Shi, 2017). Another splicing takes place in 95% of human genes, enhancing protein variety from a few genes (Baralle and Giudice, 2017; Zhu et al., 2020). Abnormal splicing is frequently observed in diseases like cancer, Duchenne muscular dystrophy, and SMA. Hence repairing of alternative splicing can be the appropriate treatment option (Scotti and Swanson, 2016; Desterro et al., 2020). Risdiplam, being the first small-molecule SMN2 splicing transformer, has been under recent human clinical trials (Ratni et al., 2018). However, the trials were stopped because of ocular toxicity in long-term animal studies (Calder et al., 2016; Ratni et al., 2016). The precursor compound of RG7800 was recognized from a phenotypic HTS. The screening depended on an SMN2 minigene luciferase assay in HEK293H cells utilizing a proprietary library (PTC Therapeutics) with higher than 200,000 small molecules (Naryshkin et al., 2014). Exon 7 inclusion improvers produce enhanced luminescence since only full-length SMN2 minigene mRNA was in accordance with the firefly luciferase sequence. The positive hits were authenticated using qRT-PCR and semiquantitative end-point RT-PCR to substantiate the raise of full-length SMN2 mRNA production. The recognized small-molecule splicing modulators (SMN-C class compounds) could augment full-length SMN2 protein levels in fibroblasts from various types of SMA patients. This was extremely selective toward the SMN2 transcript, as calculated by RNA sequencing (Naryshkin et al., 2014). These small molecules exhibited appropriate oral bioavailability and ameliorated motor function and longevity in SMA mouse models (Naryshkin et al., 2014; Feng et al., 2016; Zhao et al., 2016). NMR structures showed that SMN-C5 particularly stabilized a swollen adenine at the exon-intron junction in the helix shaped by U1 snRNA and exon 7 5′ss, thus initiating U1 snRNP detection and changing the fragile 5′ss into a robust one (Campagne et al., 2019). This promotes a novel theory for gene-specific alternative splicing repair, termed as 5′ splice site bulge repair. In addition, SMN-C5 acted jointly with the splicing dogmatic network, therefore, exhibiting more enhanced efficacy in cellular models than in vitro assays (Campagne et al., 2019). Moreover, earlier studies showed that SMN-C class compounds attached to the ESE2 region on exon 7, through substantial mechanisms applicable to SMN2 splicing activators including hnRNP G, far upstream element-binding protein 1 (FUBP1) and KH-type splicing regulatory protein (KHSRP) (Sivaramakrishnan et al., 2017; Wang et al., 2018). Nevertheless, ESE2 conjugation is not the principal cause of splicing repair, and the action mode of these RBPs have been elusive until now (Campagne et al., 2019).

Branaplam was recognized from a 1.4 million-compound library (Novartis) with a pair of contrary SMN2 minigene luciferase reporters expressed in an NSC34 MN cell line (Palacino et al., 2015). The screen showed only 1.0% positive hits. These hits exhibited remarkable enhancement of full-length SMN2 reporter luciferase signal and an associated decrease of Dexon 7 reporter signal. Depending on a pyridazine core functional group found in the hit compounds, branaplam was designed through medicinal chemistry optimization. This offered benefits in potency, bioavailability, and safety (Cheung et al., 2018). This robust drug improves full-length SMN protein generation and expands survival in a severe SMA mouse model (Palacino et al., 2015).

Conclusions and future perspectives

The primary cause of most of SMA is degeneration of MN proteins, due to the homozygous deletion or point mutation in the SMN1 gene. Unfortunately, it is the second most common genetic basis of infant mortality next to cystic fibrosis. The advent of innovative therapies in the treatment of SMA altered the natural course of the disease by altering the expression or substituting mutated genes implicated in the progression of SMA. The nematode *C. elegans* is the best model for the recognition and characterization of bioactive molecules as it helps in the quick recognition of new disease-related genetic mutants. Despite the extensive use of nusinersen currently in several nations for several types of SMA patients based primarily on inclusion criteria and economy of the country, there is a growing need to corroborate nusinersen's capability for a thorough understanding of the changeability of effect and adverse effects in a larger cohort of SMA patients. Although risdiplam is the first small-molecule SMN2 splicing transformer undergoing human clinical trials currently, the trials were ended due to ocular toxicity in long-term animal studies. Branaplam identified from a 1.4 million-compound library (Novartis) ameliorated full-length SMN protein generation and expanded survival in a severe SMA mouse model as well. In line with the currently used bioactive compounds and their protective efficacy, more studies are needed to identify novel therapeutic molecules.

References

Al-Zaidy, S.A., Kolb, S.J., Lowes, L., Alfano, L.N., Shell, R., Church, K.R., et al., 2019. AVXS-101 (onasemnogene abeparvovec) for SMA1: comparative study with a prospective natural history cohort. J. Neuromuscul. Dis. 6, 307–317.

Baek, J., Jeong, H., Ham, Y., Jo, Y.H., Choi, M., Kang, M., et al., 2019. Improvement of spinal muscular atrophy via correction of the SMN2 splicing defect by Brucea javanica (L.) Merr. extract and Bruceine, D. Phytomedicine 65, 153089.

Baralle, F.E., Giudice, J., 2017. Alternative splicing as a regulator of development and tissue identity. Nat. Rev. Mol. Cell Biol. 18, 437–451.

Burghes, A.H., Beattie, C.E., 2009. Spinal muscular atrophy: why do low levels of survival motor neuron protein make motor neurons sick? Nat. Rev. Neurosci. 10, 597–609.

Calder, A.N., Androphy, E.J., Hodgetts, K.J., 2016. Small molecules in development for the treatment of spinal muscular atrophy. J. Med. Chem. 59, 10067–10083.

Campagne, S., Boigner, S., Rudisser, S., Moursy, A., Gillioz, L., Knorlein, A., et al., 2019. Structural basis of a small molecule targeting RNA for a specific splicing correction. Nat. Chem. Biol. 15, 1191–1198.

Canuelo, A., Gilbert-Lopez, B., Pacheco-Linan, P., Martinez-Lara, E., Siles, E., Miranda-Vizuete, A., 2012. Tyrosol, a main phenol present in extra virgin olive oil, increases lifespan and stress resistance in *Caenorhabitis elegans*. Mech. Ageing Dev. 133, 563–574.

Chang, C.C., Lin, Y.T., Lu, Y.T., Liu, Y.S., Liu, J.F., 2010. Kiwifruit improves bowel function in patients with irritable bowel syndrome with constipation. Asia Pac. J. Clin. Nutr. 19, 451–457.

Cheung, A.K., Hurley, B., Kerrigan, R., Shu, L., Chin, D.N., Shen, Y., et al., 2018. Discovery of small molecule splicing modulators of survival motor neuron-2 (SMN2) for the treatment of spinal muscular atrophy (SMA). J. Med. Chem. 61, 11021–11036.

Chiriboga, C.A., 2017. Nusinersen for the treatment of spinal muscular atrophy. Expert. Rev. Neurother. 17, 955–962.

Chiriboga, C.A., Swoboda, K.J., Darras, B.T., Iannaccone, S.T., Montes, J., De Vivo, D.C., et al., 2016. Results from a phase 1 study of nusinersen (ISIS-SMN(Rx)) in children with spinal muscular atrophy. Neurology 86, 890–897.

Ciacci, C., Russo, I., Bucci, C., Iovino, P., Pellegrini, L., Giangrieco, I., et al., 2014. The kiwi fruit peptide kissper displays anti-inflammatory and anti-oxidant effects in in-vitro and ex-vivo human intestinal models. Clin. Exp. Immunol. 175, 476–484.

Ciardiello, M.A., Giangrieco, I., Tuppo, L., Tamburrini, M., Buccheri, M., Palazzo, P., et al., 2009. Influence of the natural ripening stage, cold storage, and ethylene treatment on the protein and IgE-binding profiles of green and gold kiwi fruit extracts. J. Agric. Food Chem. 57, 1565–1571.

Corey, D.R., 2017. Nusinersen, an antisense oligonucleotide drug for spinal muscular atrophy. Nat. Neurosci. 20, 497–499.

de Carlos Caceres, I., Porto, D.A., Gallotta, I., Santonicola, P., Rodriguez-Cordero, J., Di Schiavi, E., et al., 2018. Automated screening of C. elegans neurodegeneration mutants enabled by microfluidics and image analysis algorithms. Integr. Biol. 10, 539–548.

De Vivo, D.C., Bertini, E., Swoboda, K.J., Hwu, W.L., Crawford, T.O., Finkel, R.S., et al., 2019. Nusinersen initiated in infants during the presymptomatic stage of spinal muscular atrophy: interim efficacy and safety results from the Phase 2 NURTURE study. Neuromuscul. Disord. 29, 842–856.

De Wel, B., Goosens, V., Sobota, A., Van Camp, E., Geukens, E., Van Kerschaver, G., et al., 2020. Nusinersen treatment significantly improves hand grip strength, hand motor function and MRC sum scores in adult patients with spinal muscular atrophy types 3 and 4. J. Neurol. 68, 923–935.

Desterro, J., Bak-Gordon, P., Carmo-Fonseca, M., 2020. Targeting mRNA processing as an anticancer strategy. Nat. Rev. Drug Discov. 19, 112–129.

Deutsch, L., Osredkar, D., Plavec, J., Stres, B., 2021. Spinal muscular atrophy after nusinersen therapy: improved physiology in pediatric patients with no significant change in urine, serum, and liquor 1H-NMR metabolomes in comparison to an age-matched, healthy cohort. Metabolites 11, 206.

Di Giorgio, M.L., Esposito, A., Maccallini, P., Micheli, E., Bavasso, F., Gallotta, I., et al., 2017. WDR79/TCAB1 plays a conserved role in the control of locomotion and ameliorates phenotypic defects in SMA models. Neurobiol. Dis. 105, 42–50.

Di Schiavi, E., Andrenacci, D., 2013. Invertebrate models of kallmann syndrome: molecular pathogenesis and new disease genes. Curr. Genomics 14, 2–10.

Ding, A.J., Zheng, S.Q., Huang, X.B., Xing, T.K., Wu, G.S., Sun, H.Y., et al., 2017. Current perspective in the discovery of anti-aging agents from natural products. Nat. Prod. Bioprospect. 7, 335–404.

Feng, Z., Ling, K.K., Zhao, X., Zhou, C., Karp, G., Welch, E.M., et al., 2016. Pharmacologically induced mouse model of adult spinal muscular atrophy to evaluate effectiveness of therapeutics after disease onset. Hum. Mol. Genet. 25, 964–975.

Finkel, R.S., Chiriboga, C.A., Vajsar, J., Day, J.W., Montes, J., De Vivo, D.C., et al., 2016. Treatment of infantile-onset spinal muscular atrophy with nusinersen: a phase 2, open-label, dose-escalation study. Lancet 388, 3017–3026.

Fuller, H.R., Gillingwater, T.H., Wishart, T.M., 2016. Commonality amid diversity: multi-study proteomic identification of conserved disease mechanisms in spinal muscular atrophy. Neuromuscul. Disord. 26, 560–569.

Gallotta, I., Mazzarella, N., Donato, A., Esposito, A., Chaplin, J.C., Castro, S., et al., 2016. Neuron-specific knock-down of SMN1 causes neuron degeneration and death through an apoptotic mechanism. Hum. Mol. Genet. 25, 2564–2577.

Giangrieco, I., Proietti, S., Moscatello, S., Tuppo, L., Battistelli, A., La Cara, F., et al., 2016. Influence of geographical location of orchards on green kiwifruit bioactive components. J. Agric. Food Chem. 64, 9172–9179.

Gilliam, T.C., Brzustowicz, L.M., Castilla, L.H., Lehner, T., Penchaszadeh, G.K., Daniels, R.J., et al., 1990. Genetic homogeneity between acute and chronic forms of spinal muscular atrophy. Nature 345, 823–825.

Groen, E.J.N., Talbot, K., Gillingwater, T.H., 2018. Advances in therapy for spinal muscular atrophy: promises and challenges. Nat. Rev. Neurol. 14, 214–224.

Ha, J.S., Jin, D.E., Park, S.K., Park, C.H., Seung, T.W., Bae, D.W., et al., 2015. Antiamnesic effect of Actinidia arguta extract intake in a mouse model of TMT-induced learning and memory dysfunction. Evid. Based Complement Alternat. Med. 2015, 876484.

Hsu, Y.Y., Jong, Y.J., Tsai, H.H., Tseng, Y.T., An, L.M., Lo, Y.C., 2012. Triptolide increases transcript and protein levels of survival motor neurons in human SMA fibroblasts and improves survival in SMA-like mice. Br. J. Pharmacol. 166, 1114–1126.

Hua, Y., Sahashi, K., Hung, G., Rigo, F., Passini, M.A., Bennett, C.F., et al., 2010. Antisense correction of SMN2 splicing in the CNS rescues necrosis in a type III SMA mouse model. Genes Dev. 24, 1634–1644.

Huang, H.F., Guo, F., Cao, Y.Z., Shi, W., Xia, Q., 2012. Neuroprotection by manganese superoxide dismutase (MnSOD) mimics: antioxidant effect and oxidative stress regulation in acute experimental stroke. CNS Neurosci. Ther. 18, 811–818.

Illiano, P., Lanzo, A., Leo, D., Paglione, M., Zampi, G., Gainetdinov, R.R., et al., 2017. A Caenorhabditis elegans model to study dopamine transporter deficiency syndrome. Eur. J. Neurosci. 45, 207–214.

Jung, J.E., Kim, G.S., Narasimhan, P., Song, Y.S., Chan, P.H., 2009. Regulation of Mn-superoxide dismutase activity and neuroprotection by STAT3 in mice after cerebral ischemia. J. Neurosci. 29, 7003–7014.

Khan, A., Jahan, S., Imtiyaz, Z., Alshahrani, S., Antar Mankeen, S., Alshehri, BM., et al., 2020. Neuroprotection: targeting multiple pathways by naturally occurring phytochemicals. Biomedicines 8, 284.

Kolb, S.J., Coffey, C.S., Yankey, J.W., Krosschell, K., Arnold, W.D., Rutkove, S.B., et al., 2017. Natural history of infantile-onset spinal muscular atrophy. Ann. Neurol. 82, 883–891.

Kondo, T., Reaume, A.G., Huang, T.T., Carlson, E., Murakami, K., Chen, S.F., et al., 1997. Reduction of CuZn-superoxide dismutase activity exacerbates neuronal cell injury and edema formation after transient focal cerebral ischemia. J. Neurosci. 17, 4180–4189.

Kruse, T., Heller, R., Wirth, B., Gloggler, J., Wurster, C.D., Ludolph, A.C., et al., 2020. Maximum bite force in patients with spinal muscular atrophy during the first year of nusinersen therapy—a pilot study. Acta Myol. 39, 83–89.

Lefebvre, S., Bürglen, L., Reboullet, S., Clermont, O., Burlet, P., Viollet, L., et al., 1995. Identification and characterization of a spinal muscular atrophy-determining gene. Cell 80, 155–165.

Liu, J., Banskota, A.H., Critchley, A.T., Hafting, J., Prithiviraj, B., 2015. Neuroprotective effects of the cultivated Chondrus crispus in a C. elegans model of Parkinson's disease. Mar. Drugs 13, 2250–2266.

Lorson, C.L., Androphy, E.J., 2000. An exonic enhancer is required for inclusion of an essential exon in the SMA-determining gene SMN. Hum. Mol. Genet. 9, 259–265.

Lunn, M.R., Wang, C.H., 2008. Spinal muscular atrophy. Lancet 371 (9630), 2120–2133.

Luo, Y.F., Zhang, J., Liu, N.Q., Luo, Y., Zhao, B.L., 2011. Copper ions influence the toxicity of β-amyloid (1–42) in a concentration-dependent manner in a Caenorhabditis elegans model of Alzheimer's disease. Sci. China Life Sci. 54, 527–534.

Mazzarella, N., Giangrieco, I., Visone, S., Santonicola, P., Achenbach, J., Zampi, G., et al., 2019. Green kiwifruit extracts protect motor neurons from death in a spinal muscular atrophy model in Caenorhabditis elegans. Food Sci. Nutr. 7, 2327–2335.

Melki, J., 2017. Advances in spinal muscular atrophy research. In: Sumner, C.J., Paushkin, S., Ko, C. (Eds.), Spinal Muscular Atrophy—Disease Mechanisms and Therapy. Academic Press, London, UK, pp. xxiii–xxiv.

Melki, J., Abdelhak, S., Sheth, P., Bachelot, M.F., Burlet, P., Marcadet, A., et al., 1990. Gene for chronic proximal spinal muscular atrophies maps to chromosome 5q. Nature 344, 767–768.

Michelson, D., Ciafaloni, E., Ashwal, S., Lewis, E., Narayanaswami, P., Oskoui, M., et al., 2018. Evidence in focus: nusinersen use in spinal muscular atrophy: report of the guideline development, dissemination, and implementation subcommittee of the American Academy of Neurology. Neurology 91, 923–933.

Motohashi, N., Shirataki, Y., Kawase, M., Tani, S., Sakagami, H., Satoh, K., et al., 2002. Cancer prevention and therapy with kiwifruit in Chinese folklore medicine: a study of kiwifruit extracts. J. Ethnopharmacol. 81, 357–364.

Naryshkin, N.A., Weetall, M., Dakka, A., Narasimhan, J., Zhao, X., Feng, Z., et al., 2014. Motor neuron disease. SMN2 splicing modifiers improve motor function and longevity in mice with spinal muscular atrophy. Science 345, 688–693.

Osredkar, D., Jilkova, M., Butenko, T., Loboda, T., Golli, T., Fuchsova, P., et al., 2021. Children and young adults with spinal muscular atrophy treated with nusinersen. Eur. J. Paediatr. Neurol. 30, 1–8.

Palacino, J., Swalley, S.E., Song, C., Cheung, A.K., Shu, L., Zhang, X., et al., 2015. SMN2 splice modulators enhance U1-pre-mRNA association and rescue SMA mice. Nat. Chem. Biol. 11, 511–517.

Parker, J.A., Arango, M., Abderrahmane, S., Lambert, E., Tourette, C., Catoire, H., Néri, C., 2005. Resveratrol rescues mutant polyglutamine cytotoxicity in nematode and mammalian neurons. Nat. Genet. 37, 349–350.

Poirier, A., Weetall, M., Heinig, K., Bucheli, F., Schoenlein, K., Alsenz, J., et al., 2018. Risdiplam distributes and increases SMN protein in both the central nervous system and peripheral organs. Pharmacol. Res. Perspect. 6, e00447.

Polazzi, E., Mengoni, I., Caprini, M., Peña-Altamira, E., Kurtys, E., Monti, B., 2013. Copper-zinc superoxide dismutase (SOD1) is released by microglial cells and confers neuroprotection against 6-OHDA neurotoxicity. Neurosignals 21, 112–128.

Popovic, M., Andjelkovic, U., Grozdanovic, M., Aleksic, I., Gavrovic-Jankulovic, M., 2013. In vitro antibacterial activity of cysteine protease inhibitor from kiwifruit (Actinidia deliciosa). Indian J. Microbiol. 53, 100–105.

Ramdas, S., Servais, L., 2020. New treatments in spinal muscular atrophy: an overview of currently available data. Expert. Opin. Pharmacother. 21, 307–315.

Ratni, H., Karp, G.M., Weetall, M., Naryshkin, N.A., Paushkin, S.V., Chen, K.S., et al., 2016. Specific correction of alternative survival motor neuron 2 splicing by small molecules: discovery of a potential novel medicine to treat spinal muscular atrophy. J. Med. Chem. 59, 6086–6100.

Ratni, H., Ebeling, M., Baird, J., Bendels, S., Bylund, J., Chen, K.S., et al., 2018. Discovery of risdiplam, a selective survival of motor neuron-2 (SMN2) gene splicing modifier for the treatment of spinal muscular atrophy (SMA). J. Med. Chem. 61, 6501–6517.

Recio-Rodriguez, J.I., Gomez-Marcos, M.A., Patino-Alonso, M.C., Puigdomenech, E., Notario-Pacheco, B., Mendizabal-Gallastegui, N., et al., 2015. Effects of kiwi consumption on plasma lipids, fibrinogen and insulin resistance in the context of a normal diet. Nutr. J. 14, 97.

Rigo, F., Hua, Y., Krainer, A.R., Bennett, C.F., 2012. Antisense-based therapy for the treatment of spinal muscular atrophy. J. Cell Biol. 199.

Scotti, M.M., Swanson, M.S., 2016. RNA mis-splicing in disease. Nat. Rev. Genet. 17, 19–32.

Shi, Y., 2017. Mechanistic insights into precursor messenger RNA splicing by the spliceosome. Nat. Rev. Mol. Cell Biol. 18, 655–670.

Shukla, V., Phulara, S.C., Yadav, D., Tiwari, S., Kaur, S., Gupta, M.M., et al., 2012. Iridoid compound 10-O-trans-p-coumaroylcatalpol extends longevity and reduces α synuclein aggregation in Caenorhabditis elegans. CNS Neurol. Disord. Drug Targets 11, 984–992.

Simon, C.M., Dai, Y.A., Van Alstyne, M., Koutsioumpa, C., Pagiazitis, J.G., Chalif, J.I., et al., 2017. Converging mechanisms of p53 activation drive motor neuron degeneration in spinal muscular atrophy. Cell Rep. 21, 3767–3780.

Singh, N.K., Singh, N.N., Androphy, E.J., Singh, R.N., 2006. Splicing of a critical exon of human survival motor neuron is regulated by a unique silencer element located in the last intron. Mol. Cell. Biol. 26, 1333–1346.

Singh, R.N., Howell, M.D., Ottesen, E.W., Singh, N.N., 2017. Diverse role of survival motor neuron protein. Biochim. Biophys. Acta Gene. Regul. Mech. 1860, 299–315.

Sivaramakrishnan, M., McCarthy, K.D., Campagne, S., Huber, S., Meier, S., Augustin, A., et al., 2017. Binding to SMN2 pre-mRNA-protein complex elicits specificity for small molecule splicing modifiers. Nat. Commun. 8, 1476.

Srivastava, S., Sammi, S.R., Laxman, T.S., Pant, A., Nagar, A., Trivedi, S., et al., 2017. Silymarin promotes longevity and alleviates Parkinson's associated pathologies in Caenorhabditis elegans. J. Funct. Foods 31, 32–43.

Stepek, G., Buttle, D.J., Duce, I.R., Lowe, A., Behnke, J.M., 2005. Assessment of the anthelmintic effect of natural plant cysteine proteinases against the gastrointestinal nematode, Heligmosomoides polygyrus, in vitro. Parasitology 130, 203–211.

Stolte, B., Totzeck, A., Kizina, K., Bolz, S., Pietruck, L., Monninghoff, C., et al., 2018. Feasibility and safety of intrathecal treatment with nusinersen in adult patients with spinal muscular atrophy. Ther. Adv. Neurol. Disord. 11. 1756286418803246.

Sugarman, E.A., Nagan, N., Zhu, H., Akmaev, V.R., Zhou, Z., Rohlfs, E.M., et al., 2012. Pan-ethnic carrier screening and prenatal diagnosis for spinal muscular atrophy: clinical laboratory analysis of >72,400 specimens. Eur. J. Hum. Genet. 20, 27–32.

Wadman, R.I., Stam, M., Gijzen, M., Lemmink, H.H., Snoeck, I.N., Wijngaarde, C.A., et al., 2017. Association of motor milestones, SMN2 copy and outcome in spinal muscular atrophy types 0–4. J. Neurol. Neurosurg. Psychiatry 88, 365–367.

Wang, J., Schultz, P.G., Johnson, K.A., 2018. Mechanistic studies of a small-molecule modulator of SMN2 splicing. Proc. Natl. Acad. Sci. U. S. A. 115, E4604–E4612.

Xue, W.Z., Yang, Q.Q., Chen, Y., Zou, R.X., Xing, D., Xu, Y., et al., 2017. Kiwifruit alleviates learning and memory deficits induced by Pb through antioxidation and inhibition of microglia activation in vitro and in vivo. Oxid. Med. Cell. Longev. 2017, 5645324.

Yeo, C.J.J., Darras, B.T., 2020. Overturning the paradigm of spinal muscular atrophy as just a motor neuron disease. Pediatr. Neurol. 109, 12–19.

Young, P.J., Le, T.T., thi Man, N., Burghes, A.H.M., Morris, G.E., 2000. The relationship between SMN, the spinal muscular atrophy protein, and nuclear coiled bodies in differentiated tissues and cultured cells. Exp. Cell Res. 256, 365–374.

Zhao, X., Feng, Z., Ling, K.K., Mollin, A., Sheedy, J., Yeh, S., et al., 2016. Pharmacokinetics, pharmacodynamics and efficacy of a small-molecule SMN2 splicing modifier in mouse models of spinal muscular atrophy. Hum. Mol. Genet. 25, 1885–1899.
Zhu, S., Rooney, S., Michlewski, G., 2020. RNA-targeted therapies and high-throughput screening methods. Int. J. Mol. Sci. 21, 2996.

CHAPTER 8

Improvement of amyotrophic lateral sclerosis symptoms using plant extracts

Introduction

Pathology

Jean-Martin Charcot explained amyotrophic lateral sclerosis (ALS) for the first time in the 19th century when he observed particular patterns of spinal cord white and gray matter damage in patients experiencing muscle weakness and atrophy (Goetz, 2000). The lifespan of patients after diagnosis is generally 2–3 years since they undergo respiratory failure due to the weakness in respiratory muscles (Zarei et al., 2015). A wealth of studies from the 1990s on biochemistry, genetics, and pathology reported that ALS and frontotemporal dementia (FTD) share similar pathological events (Neary et al., 1990; Mitsuyama, 1993: Novak et al., 2021). Riluzole and edaravone, the presently used drugs, show only a moderate amelioration in lifespan (Yoshino and Kimura, 2006; Miller et al., 2002). Since most ALS patients lack a family history of disease, they are regarded as sporadic ALS (sALS) patients (Tandan and Bradley, 1985); around 5% report a family history of the disease because of inheritance of a mutation in an ALS-linked gene and are regarded as familial ALS (fALS) patients (Byrne et al., 2011; Kok et al., 2021).

Several studies on fALS patients discovered numerous genes that are responsibly linked to ALS, of which the chromosome 9 open reading frame 72 (*C9ORF72*) gene (DeJesus-Hernandez et al., 2011; Renton et al., 2011) and the copper-zinc superoxide dismutase (*SOD1*) gene are common (Rosen et al., 1993). Studies on fALS genes have indicated numerous probable mechanisms of motor neuron degeneration such as excitotoxicity, oxidative stress, protein aggregation, and shortcomings in cell pathways like autophagy, RNA metabolism, and the DNA damage response (DDR) (Ferraiuolo et al., 2011). A wealth of studies has shown that fALS genes are involved in the DDR, and DNA damage has been the underlying cause

Plant Extracts in Neurodegenerative Diseases
https://doi.org/10.1016/B978-0-323-95762-5.00005-9

of both sporadic and familial ALS since the 1990s (Ferrante et al., 1997; Fitzmaurice et al., 1996), earlier than several of these genes associated to ALS. DNA damage and deficit in the DDR, therefore, probably has an essential role in ALS.

DNA damage

DNA damage takes place in cells with an approximation of 10^4–10^5 DNA lesions per day. Lack of the repair process induces deleterious effects eventually leading to cell death (Giglia-Mari et al., 2011). The probability of DNA damage is accidental at the time of transcription or because of exposure to hazardous genotoxic agents that can show an impact both on nuclear and mitochondrial DNA (Kazak et al., 2012). For example, DNA damage involves deamination of bases or oxidation; base insertions, deletions, or substitutions; and DNA double- or single-stranded breaks (DSBs or SSBs).

DNA damage and DNA repair deficit have been linked to neurodegeneration as mutations in DDR and DNA repair genes have been observed to induce several neurodegenerative diseases (NDs) such as ataxia telangiectasia, ataxia telangiectasia-like disorder, Nijmegen breakage syndrome, ataxia-oculomotor apraxia-1 and 2, ALS4 (instigated by mutations in *SETX*, (Chen et al., 2004)), spinocerebellar ataxia with axonal neuropathy-1, and Cockayne syndrome (Jeppesen et al., 2011).

Especially, the numerous NDs induced by DDR or DNA repair gene mutations particularly perturb cerebellar neurons instead of motor neurons. Motor neurons can also be influenced by DNA repair deficit as mice with decreased expression of Ercc1, a protein responsible for nucleotide excision repair (NER), show age-dependent motor neuron degeneration and astrogliosis, in line with ALS (De Waard et al., 2010).

First intron of chromosome 9 open reading frame 72 possesses a hexanucleotide (GGG GCC) repeat expansion that plays a vital role in the etiopathogenesis of fALS in the West (DeJesus-Hernandez et al., 2011; Renton et al., 2011). Voluminous data has discovered enhanced protein expression and staining for γH2AX in *C9ORF72*-ALS patient postmortem spinal cord tissue and induced pluripotent stem cells (iPSC)-obtained motor neurons, indicating DNA damage is accrued in *C9ORF72*-ALS (Lopez-Gonzalez et al., 2016; Farg et al., 2017; Walker et al., 2017; Higelin et al., 2018; Andrade et al., 2020).

Viral expression of *C9ORF72*-ALS RNA repeat expansion (RREs) or a few DPRs in neuronal cells provoke DNA strand breaks and augmented γH2AX levels (Lopez-Gonzalez et al., 2016; Farg et al., 2017; Nihei et al.,

2020), indicating that DNA damage in *C9ORF72*-ALS is initiated by RREs or DPRs. ROS enhanced by mitochondrial impairment was instigated by poly(GR) expression damage DNA (Choi et al., 2019). Likewise, R-loops that, in general, exist as RNA:DNA hybrids and promote DNA strand breaks are enhanced in C9ORF72-ALS postmortem spinal cord tissue and poly(GA) DPR-transfected cells (Walker et al., 2017).

Additional importance of the repeat expansion in C9ORF72-ALS is the collection of protein-associated DNA breaks induced by surrounded TOP1 cleavage complexes (TOP1cc) as noticed in cells expressing RREs and DPRs (Walker et al., 2017; Walker and El-Khamisy, 2018). Moreover, these mechanisms probably overlap each other since ROS induce both R-loops and TOP1cc (Cristini et al., 2019). Additionally, ROS-mediated aggregation of cotranscriptional R-loops and TOP1cc-dependent SSBs in reverse neighboring DNA strands promotes DSBs through transcription (Sordet et al., 2009; Alagoz et al., 2013; Teng et al., 2018; Cristini et al., 2019), therefore worsening a cellular injury that can eventually lead to neuronal death.

DDR signaling and DNA repair have been found to be impaired in *C9ORF72*-ALS and could augment DNA damage. Repeat expansions have also been observed to hamper DNA repair (Usdin et al., 2015). Additionally, chromatin condensation and expression of factors that initiate this process are enhanced in *C9ORF72*-ALS spinal cord tissue and poly(GA) transfected cells, which could hamper contact to DNA repair factors. In line with this, instigating chromatin relaxation mitigated γH2AX levels and cell death in poly(GA)-transfected cells (Walker et al., 2017).

Interestingly, p53 seems to have a role in *C9ORF72*-ALS and DPR-promoted toxicity as p53 knockout or knockdown has been found to improve the lifespan of a mouse model with poly(PR) expression and rescue from neurodegeneration in *Drosophila* models expressing the *C9ORF72* repeat expansion (Maor-Nof et al., 2021). Amazingly, p53 knockout also curtails DNA damage (γH2AX levels and comet tail measurements) in poly(PR)-transduced cells and *C9ORF72*-ALS iPSC-obtained motor neurons, suggesting p53 action may be taking place upstream of DNA damage instead of downstream (Maor-Nof et al., 2021).

Plant extracts

Natural products derived from plant extracts exhibit exemplary protection against oxidative stress (Pohl and Thoo Lin, 2018). There has been a smaller number of drugs getting approval for NDs within the last 8 years. Only one new drug was discovered for ALS (Jackson et al., 2019) and HD (Austedo, CenterWatch, 2018). In pharmaceutical industries, a plethora of plant natural products (single compounds extracted from natural sources) and/or plant extracts (comprising multifarious secondary metabolites) have exerted antioxidant activity. Plant metabolites have offered motivation for medicinal chemists and a platform for drug development processes for many years (Newman and Cragg, 2016). Unfortunately, within the last 2 decades, the use of plant metabolites in drug discovery has been mitigated, primarily because of the technical snags for screening natural products in high-throughput assays for particular molecular targets, biological riddles linked to the synthesis of natural compounds (Harvey et al., 2015), progress in metagenomics, and the development of combinational chemistry (Mathur and Hoskins, 2017). Nevertheless, currently there has been a repetition of penetrating interest in plant metabolites. This is because of the progression in enhanced fractionation and advent of novel NMR techniques for structural understanding. This also involves profiling and isolation techniques like HPLC-MS/MS, high-resolution Fourier-transform mass spectrometry, and photo-diode arrays for metabolomics (Harvey et al., 2015).

Despite the disease-defending characteristics of a few secondary plant metabolites in vivo, they are nonessential compounds for humans. A plethora of plant phytochemicals from plants exert a wide range of biological activities including antiinflammatory (Harbourne et al., 2013; Khlifi and Sghaier, 2013; Muthaiyah et al., 2011; Saeidnia and Gohari, 2012), antimicrobial (Wink and Abbas, 2013; Yue et al., 2014), anticarcinogenic (Sun-Waterhouse, 2011; Wink and Abbas, 2013), and antidiabetic (Fontana et al., 2013; Brusotti et al., 2014) properties. A few of these activities are because of controlling of the molecular pathways (Upadhyay and Dixit, 2015). With these beneficial effects, they are considered of significant value in human health. In line with this, the in vitro and in vivo effects of phytochemicals have exhibited considerable interest as reported in several studies, with their efficacy to treat and avert neurodegeneration (Björkman et al., 2011; Fontana et al., 2013; Lahlou, 2013; Yue et al., 2014).

With a view to investigating NDs and the efficacy of natural plant products or plant extracts for their inhibition or treatment, diverse model organisms including rodents (mice, rats), fruit fly (*Drosophila melanogaster*), zebrafish (*Danio rerio*), nematodes (*Caenorhabditis elegans*), and unicellular organisms like yeast (*Saccharomyces cerevisiae*) have been explored (Gitler et al., 2017). In addition, cell-based studies are also used to evaluate the neurodegeneration process and efficacy of treatment. Presently, studies have produced induced pluripotent stem cells that undergo division and self-renew, thus maturing into all cells of the adult body by differentiated patient-obtained cells (Bahmad et al., 2017). Further, development of 3D models with the help of human stem cells has also been made possible (Marton and Pasca, 2016; Gitler et al., 2017; Jorfi et al., 2018).

Clinical trials

Recently, an herbal drug, altered jakyakgamchobuja-tang, was administered to ALS patients. The study reported that 3 months of collective therapy using herbal drug, acupuncture, pharmacopuncture, and needle-embedding therapy profoundly mitigated ALS advancement. Accordingly, the Korean type of the ALS Functional Rating Scale-Revised (K-ALSFRS-R) score enhanced from 28.42 ± 7.83 to 29.08 ± 7.99 in a month's time from the beginning of the treatment and reduced to 28.16 ± 8.23 after 3 months. Nevertheless, it drastically reduced to 21.33 ± 9.93 after 8 months from discharge (Kim et al., 2010). Other researchers reported that 30 days of the collective treatment resulted in a K-ALSFRS-R score augmentation from 28.42 ± 7.83 to 29.08 ± 7.99 (Kim et al., 2009). Mecasin, also called KCHO-1, an herbal extract from altered jakyakgamchobuja-tang, has exerted neuroprotective and antineuroinflammatory efficacy and showed no toxicity in both in vitro and in vivo trials (Lee et al., 2014, 2016a; Cha et al., 2015a,b; Jeong et al., 2014; Lee et al., 2016b; Kook et al., 2017).

Mecasin

Mecasin, also familiar as KCHO-1, is prepared by using ethanolic extracts of nine medicinal herbs: *Curcuma longa, Salvia miltiorrhiza, Gastrodia elata, Chaenomeles sinensis, Polygala tenuifolia, Paeonia lactiflora, Glycyrrhiza uralensis, Atractylodes lancea,* and treated *Aconitum carmichaeli*. Mecasin is prepared by Hanpoong Pharm and Foods Co. (Chonju, South Korea) with well-established manufacturing practices by the Ministry of Food and Drug Safety in the form of a rectangular maroon tablet enclosed in a film coating. Although animal studies have shown the protective effect at doses of between 250 and 500 mg/kg/day (Jeong et al., 2014; Cha et al., 2015a), toxicological studies showed no toxicity at >5000 mg/kg/day.

Ashwagandha

Ashwagandha extract was introduced orally to hTDP-43A315T mice for 8 weeks beginning from 64 to 48 weeks of age for males and females, individually. Interestingly, hTDP-43A315T mice showed improvement in their motor performance on rotarod test and cognitive activity evaluated by the passive avoidance test. Microscopy examination of tissue samples showed that Ashwagandha treatment of hTDP-43A315T mice ameliorated innervation at neuromuscular junctions, suppressed neuroinflammation, and curtailed NF-κB activation (Dutta et al., 2017). Curiously, Ashwagandha treatment inverted the cytoplasmic displacement of hTDP-43 in spinal motor neurons and in brain cortical neurons of hTDP-43A315T mice, and it mitigated hTDP-43 accumulation. Mounting evidence in vitro has shown that the neuronal protection of TDP-43 displacement is probably because of the incidental effect of factors liberated from microglial cells treated with Ashwagandha. Based on these findings, it can be concluded that Ashwagandha and its components might signify capable therapeutics for TDP-43 proteinopathies (Dutta et al., 2017).

Withaferin A

Transgenic mice expressing human TDP-43A315T mutant gene when treated with Withaferin A, an NF-κB suppressor, improved disease phenotypes. Withaferin A is a constituent of the plant *Withania somnifera dunal* [or Ashwagandha (ASH)], a perennial plant of family Solanaceae (Singh et al., 2011) used for 4000 years in the traditional Indian medical system (Ayurveda). The root of the plant harbors 14–15 multifarious alkaloids and around 40 structurally comparable steroidal lactones (called withanolides) besides different kinds of carbohydrates and amino acids, in variable quantities (Upton et al., 2000). It shows antimicrobial, antiinflammatory, antineoplastic, antistress, cardioprotective, antidiabetic, and neuroprotective properties (Mishra et al., 2000). The neuroprotective activity of the plant extract or purified components from the plant has been explored in a plethora of disease models including Alzheimer's disease, Parkinson's disease, Huntington's disease, and spinal cord injuries (Kuboyama et al., 2014).

Studies on microglial cultured BV2 cells showed that an extract of *W. somnifera* (ASH) substantially curtailed the activity of NF-κB P65–luciferase reporter in a dose-dependent manner. The antiinflammatory activity of ASH was also shown by its efficacy in attenuating proinflammatory cytokines/chemokines discharged by primary mouse microglia in response to an inflammatory stimulus. Of the various types of cytokines and chemokines perturbed, a few with accrued circulating levels in serum and/or plasma such as RANTES, IL-6, IL-17, and MIP-1α drew attention (Rentzos et al., 2007, 2010; Ono et al., 2001; Yang et al., 2016). Similarly, soluble TNF receptors R1 and R2 levels are also observed to be enhanced in plasma of patients with ALS (Cereda et al., 2008). IL-4 levels have been noticed to be undesirably interrelated with IL-6 levels in cerebrospinal fluid (CSF) of ALS cases. Therefore, if IL-6 level enhances, IL-4 reduces (Martinez et al., 2016). Peripherin, an intermediate filament protein aggregation, is a pathological event of importance in ALS, and transgenic mouse studies showed that continued peripherin overexpression induces motor neuron death (Beaulieu et al., 1999, 2000). Surprisingly, peripherin levels are upregulated in the hTDP-43A315T transgenic mice (Swarup et al., 2011) and were curtailed by ASH treatment.

In line with the in vitro outcome, the protective efficacy of ASH on neuronal pathology appears to indirectly evoke from the effect of the drug on glial cells. Essentially, it is understood that glial cells play a crucial role in the pathogenesis of ALS (Philips and Rothstein, 2014). Astrocytes have been observed to result in motor neuron death in a plethora of studies (Maragakis and Rothstein, 2006; Haidet-Phillips et al., 2011). However, astrocyte initiation by factors liberated from impaired motor neurons probably provoke the release of nerve growth factor, and along with other antioxidative enzymes render neuroprotection (Pehar et al., 2005). Microglia also play a multifaceted role. Regarding neurodegeneration or aggregations of misfolded proteins, microglia can multiply and assume a triggered state with inflow of numerous factors that probably are advantageous or harmful to neurons (Norden et al., 2015; Perry and Holmes, 2014). Therefore, microglial responses can differ from neuroprotective (M2a state) to neurodegenerative/toxic states (M1, classically triggered) (Chhor et al., 2013). In a recent study, 8 weeks of ASH treatment showed a profound decrease in GFAP expression in the spinal cord of hTDP-43A315T mice and decrease of Iba-1 immunoreactivity in microglia, despite these numerous cells' inability to undertake a quiescent morphology after ASH treatment.

Drawbacks

Although ASH has shown beneficial neuroprotective effects, a few drawbacks limit its therapeutic approach. Similar to a few plant extracts, it is unclear which ingredients of ASH induce which effect and if the protection of compounds acting alone or cooperatively with others in whole extracts. Furthermore, enormous differences in composition of ASH extracts may arise due to the differences in the source of *W. somnifera* or methods of root extract preparations.

Polyphenols

Polyphenols are natural compounds with numerous substantial neuroprotective effects as shown in a wide range of models of NDs like Alzheimer's and Parkinson's disease. They are primarily used as dietary sources and medicines to combat NDs and act as therapeutics to deliberate disease development and mitigate symptoms (Silva and Pogacnik, 2020; Obulesu, 2021). They are a varied group of naturally available compounds having a chemical structure with one or more phenolic rings. They are abundantly available in plant foods including fruits, vegetables, and whole grains (Swallah et al., 2020; Rambaran, 2020). Polyphenols are characterized as secondary metabolites and play an essential role in normal growth and in the plant rescue system (Yang et al., 2020). They are prepared in the shikimate and phenylpropanoid pathways (Tsao, 2010). A plethora of polyphenols have been identified to exert neuroprotective action in mammalian cell and animal models of ALS/FTD (Solanki et al., 2015).

Resveratrol

Protective efficacy of resveratrol in ALS were first investigated in neuronal cell lines expressing the SOD1G93A mutant gene (Kim et al., 2007; Barber et al., 2009; Wang et al., 2011). Interestingly, it reduced cell death by 50%, which was noticed as a result of SOD1-mediated toxicity (Kim et al., 2007). Treatments of mouse motor neuron cells NSC34 expressing SOD1G93A exhibited a little dose-dependent amelioration in cell viability and a parallel decrease in the cytosolic ROS concentration (Barber et al., 2009). Resveratrol treatment rescued rat cortical motor neurons from the adverse effects of CSF from ALS patients (Yáñez et al., 2011). Conversely, studies in mice ALS models expressing mutant SOD1G93A demonstrated contradictory results, which may be due to various procedures on dosing and route of administration. Continuing oral introduction of resveratrol at 25 mg/kg/day failed to ameliorate motor functions and lifespan of ALS mice (Markert et al., 2010). On the contrary, intraperitoneal injections of 20 mg/kg/twice a week ameliorated survival and deferred the onset of ALS (Han et al., 2012). The analogous encouraging effect on survival and motor function was noticed with an increased dose (160 mg/kg/day) introduced orally (Mancuso et al., 2014a). The neuroprotective efficacy of resveratrol in ALS mice has been exhibited in coadministration with other robust therapeutics (Mancuso et al., 2014a; Schiaffino et al., 2018).

Curcumin

The substantial protective effects of curcumin were studied in numerous NDs. In models of Alzheimer's and Parkinson's disease, curcumin can curtail oxidative stress, inhibit toxic protein accumulation, and rescue against apoptosis (Adami and Bottai, 2020; Soo et al., 2020). In ALS, curcumin was found to hamper accumulation of reduced weight SOD1 in vitro by attaching to its aggregation-provoked regions. Curcumin-bound SOD1 aggregates were smaller, without structure, and showed reduced cytotoxicity (Bhatia et al., 2015).

Conclusions and future perspectives

Although there is a potential link between DNA damage and neurodegeneration, there are no adequate studies on the DDR and DNA repair in neurons. In addition, it is hitherto unknown if particular motor neuron types are highly susceptible to DNA damage or DNA repair deficit. When DNA repair kinetics were correlated between astrocytes and neurons, it was understood that both astrocytes and neurons are based on TDP1 for DNA SSB repair (Katyal et al., 2007).

Presently, due to the lack of an everlasting therapy for ALS with conventional medicine (Aronson, 2006), patients probably focus on a multidisciplinary clinic or opt for a complementary and alternative medicine (Stephens et al., 2016; Kim et al., 2016). After the rigorous research of neuroscientists, merely two treatment options with a little efficacy on survival and motor function are presently in use for ALS. The manifold compounds studied as probable ALS therapies hitherto were observed to be futile in clinical trials, thus accentuating the necessity for further research (Petrov et al., 2017; Novak et al., 2021).

References

Adami, R., Bottai, D., 2020. Curcumin and neurological diseases. Nutr. Neurosci. 25, 1–21.

Alagoz, M., Chiang, S., Sharma, A., El-Khamisy, S.F., 2013. ATM deficiency results in accumulation of DNA-topoisomerase I covalent intermediates in neural cells. PLoS One 8, e58239.

Andrade, N.S., Ramic, M., Esanov, R., Liu, W., Rybin, M.J., Gaidosh, G., et al., 2020. Dipeptide repeat proteins inhibit homology-directed DNA double strand break repair in C9ORF72 ALS/FTD. Mol. Neurodegener. 15, 1–18.

Aronson, J.K., 2006. Rare diseases and orphan drugs. Br. J. Clin. Pharmacol. 61, 243–245.

Bahmad, H., Hadadeh, O., Chamaa, F., Cheaito, K., Darwish, B., Makkawi, A.K., et al., 2017. Modeling human neurological and neurodegenerative diseases: from induced pluripotent stem cells to neuronal differentiation and its applications in neurotrauma. Front. Mol. Neurosci. 10, 1–17.

Barber, S.C., Higginbottom, A., Mead, R.J., Barber, S., Shaw, P.J., 2009. An in vitro screening cascade to identify neuroprotective antioxidants in ALS. Free Radic. Biol. Med. 46, 1127–1138.

Beaulieu, J.M., Nguyen, M.D., Julien, J.P., 1999. Late onset of motor neurons in mice overexpressing wild-type peripherin. J. Cell Biol. 147, 531–544.

Beaulieu, J.M., Jacomy, H., Julien, J.P., 2000. Formation of intermediate filament protein aggregates with disparate effects in two transgenic mouse models lacking the neurofilament light subunit. J. Neurosci. 20, 5321–5328.

Bhatia, N.K., Srivastava, A., Katyal, N., Jain, N., Khan, M.A.I., Kundu, B., et al., 2015. Curcumin binds to the pre-fibrillar aggregates of Cu/Zn superoxide dismutase (SOD1) and alters its amyloidogenic pathway resulting in reduced cytotoxicity. Biochim. Biophys. Acta, Proteins Proteomics 1854, 426–436.

Björkman, M., Klingen, I., Birch, A.N.E., Bones, A.M., Bruce, T.J.A., Johansen, T.J., et al., 2011. Phytochemicals of Brassicaceae in plant protection and human health—influences of climate, environment and agronomic practice. Phytochemistry 72, 538–556.

Brusotti, G., Cesari, I., Dentamaro, A., Caccialanza, G., Massolini, G., 2014. Isolation and characterization of bioactive compounds from plant resources: the role of analysis in the ethnopharmacological approach. J. Pharm. Biomed. Anal. 87, 218–228.

Byrne, S., Walsh, C., Lynch, C., Bede, P., Elamin, M., Kenna, K., et al., 2011. Rate of familial amyotrophic lateral sclerosis: a systematic review and meta-analysis. J. Neurol. Neurosurg. Psychiatry 82, 623–627.

CenterWatch, 2018. FDA Approved Drugs by Medical Condition. Available from: https://www.centerwatch.com/drug-information/fda-approved-drugs/medical-conditions/. (Accessed 8 November 2018).

Cereda, C., Baiocchi, C., Bongioanni, P., Cova, E., Guareschi, S., Metelli, M.R., et al., 2008. TNF and sTNFR1/2 plasma levels in ALS patients. J. Neuroimmunol. 194, 123–131.

Cha, E., Jeong, H., Lee, J., Lee, S., Park, M., Kim, S., 2015a. A study on single dose toxicity of Mecasin pharmacopuncture injection in muscle. J. Korean Med. 36, 36–42.

Cha, E., Lee, J., Lee, S., Park, M., Song, I., Son, I., et al., 2015b. A 4-week repeated dose oral toxicity study of Mecasin in Sprague-Dawley rats to determine the appropriate doses for a 13-week, repeated toxicity test. Aust. J. Pharm. 18, 45–50.

Chen, Y.Z., Bennett, C.L., Huynh, H.M., Blair, I.P., Puls, I., Irobi, J., et al., 2004. DNA/RNA helicase gene mutations in a form of juvenile amyotrophic lateral sclerosis (ALS4). Am. J. Hum. Genet. 74, 1128–1135.

Chhor, V., Le Charpentier, T., Lebon, S., Ore, M., Celador, I.L., Josserand, J., et al., 2013. Characterization of phenotype markers and neuronotoxic potential of polarised primary microglia in vitro. Brain Behav. Immun. 32, 70–85.

Choi, S.Y., Lopez-Gonzalez, R., Krishnan, G., Phillips, H.L., Li, A.N., Seeley, W.W., et al., 2019. C9ORF72-ALS/FTD-associated poly(GR) binds Atp5a1 and compromises mitochondrial function in vivo. Nat. Neurosci. 22, 851–862.

Cristini, A., Ricci, G., Capranico, G., Salimbeni, S., Huang, S.Y., Marinello, J., et al., 2019. Dual processing of R-loops and topoisomerase I induces transcription-dependent DNA double-strand breaks. Cell Rep. 28, 3167–3181.

De Waard, M.C., Van Der Pluijm, I., Zuiderveen Borgesius, N., Comley, L.H., Haasdijk, E.D., Rijksen, Y., et al., 2010. Age-related motor neuron degeneration in DNA repair deficient Ercc1 mice. Acta Neuropathol. 120, 461–475.

DeJesus-Hernandez, M., Mackenzie, I.R., Boeve, B.F., Boxer, A.L., Baker, M., Rutherford, N.J., et al., 2011. Expanded GGG GCC hexanucleotide repeat in noncoding region of C9ORF72 causes chromosome 9p-linked FTD and ALS. Neuron 72, 245–256.

Dutta, K., Patel, P., Rahimian, R., Phaneuf, D., Julien, J., 2017. Withania somnifera reverses transactive response DNA binding protein 43 proteinopathy in a mouse model of amyotrophic lateral sclerosis/frontotemporal lobar degeneration. Neurotherapeutics 14, 447–462.

Farg, M.A., Konopka, A., Soo, K.Y., Ito, D., Atkin, J.D., 2017. The DNA damage response (DDR) is induced by the C9orf72 repeat expansion in amyotrophic lateral sclerosis. Hum. Mol. Genet. 26, 2882–2896.

Ferraiuolo, L., Kirby, J., Grierson, A.J., Sendtner, M., Shaw, P.J., 2011. Molecular pathways of motor neuron injury in amyotrophic lateral sclerosis. Nat. Rev. Neurol. 7, 616–630.

Ferrante, R.J., Browne, S.E., Shinobu, L.A., Bowling, A.C., Baik, M.J., MacGarvy, U., et al., 1997. Evidence of increased oxidative damage in both sporadic and familial amyotrophic lateral sclerosis. J. Neurochem. 69, 2064–2074.

Fitzmaurice, P.S., Shaw, I.C., Kleiner, H.E., Miller, R.T., Monks, T.J., Lau, S.S., et al., 1996. Evidence for DNA damage in amyotrophic lateral sclerosis. Muscle Nerve 19, 797–798.

Fontana, A.R., Antoniolli, A., Bottini, R., 2013. Grape pomace as a sustainable source of bioactive compounds: extraction, characterization, and biotechnological applications of phenolics. J. Agric. Food Chem. 61, 8987–9003.

Giglia-Mari, G., Zotter, A., Vermeulen, W., 2011. DNA damage response. Cold Spring Harb. Perspect. Biol. 3, a000745.

Gitler, A.D., Dhillon, P., Shorter, J., 2017. Neurodegenerative disease: models, mechanisms, and a new hope. Dis. Model. Mech. 10, 499–502.

Goetz, C.G., 2000. Amyotrophic lateral sclerosis: early contributions of Jean Martin Charcot. Muscle Nerve 23, 336–343.

Haidet-Phillips, A.M., Hester, M.E., Miranda, C.J., Meyer, K., Braun, L., Frakes, A., et al., 2011. Astrocytes from familial and sporadic ALS patients are toxic to motor neurons. Nat. Biotechnol. 29, 824–828.

Han, S., Choi, J.R., Soon Shin, K., Kang, S.J., 2012. Resveratrol upregulated heat shock proteins and extended the survival of G93A-SOD1 mice. Brain Res. 1483, 112–117.

Harbourne, N., Marete, E., Jacquier, J.C., O'Riordan, D., 2013. Stability of phytochemicals as sources of anti-inflammatory nutraceuticals in beverages—a review. Food Res. Int. 50, 480–486.

Harvey, A.L., Edrada-Ebel, R., Quinn, R.J., 2015. The re-emergence of natural products for drug discovery in the genomics era. Nat. Rev. Drug Discov. 14, 111–129.

Higelin, J., Catanese, A., Semelink-Sedlacek, L.L., Oeztuerk, S., Lutz, A., Bausinger, J., et al., 2018. NEK1 loss-of-function mutation induces DNA damage accumulation in ALS patient-derived motoneurons. Stem Cell Res. 30, 150–162.

Jackson, C., Patterson, T.H., Kittrell, P., Baranovsky, T., MCAnanama, G., Bower, L., et al., 2019. Radicava (edaravone) for amyotrophic lateral sclerosis: US experience at 1 year after launch. Amyotroph. Lateral Scler. Front. Degener. 7-8, 605–610.

Jeong, H., Lee, J., Cha, E., Park, M., Son, I., Song, B., et al., 2014. A study on the oral toxicity of Mecasin in rats. Aust. J. Pharm. 17, 61.

Jeppesen, D.K., Bohr, V.A., Stevnsner, T., 2011. DNA repair deficiency in neurodegeneration. Prog. Neurobiol. 94, 166–200.

Jorfi, M., D'Avanzo, C., Kim, D.Y., Irimia, D., 2018. Three-dimensional models of the human brain development and diseases. Adv. Healthc. Mater. 7, 1–20.

Katyal, S., El-Khamisy, S.F., Russell, H.R., Li, Y., Ju, L., Caldecott, K.W., et al., 2007. TDP1 facilitates chromosomal single-strand break repair in neurons and is neuroprotective in vivo. EMBO J. 26, 4720–4731.

Kazak, L., Reyes, A., Holt, I.J., 2012. Minimizing the damage: repair pathways keep mitochondrial DNA intact. Nat. Rev. Mol. Cell Biol. 13, 659–671.

Khlifi, D., Sghaier, R.M., 2013. Anti-inflammatory and acetylcholinesterase inhibition activities of Globularia Alypum. J. Med. Bioeng. 2, 232–237.

Kim, D., Nguyen, M.D., Dobbin, M.M., Fischer, A., Sananbenesi, F., Rodgers, J.T., et al., 2007. SIRT1 deacetylase protects against neurodegeneration in models for Alzheimer's disease and amyotrophic lateral sclerosis. EMBO J. 26, 3169–3179.

Kim, S., Na, W., Lim, N., Lee, D., Jang, E., Song, B., 2009. A pilot clinical study on the traditional Korean medicine treatment of amyotrophic lateral sclerosis. J. Pharmacopuncture 12, 53–65.

Kim, H., Song, B., Park, M., Lim, N., Kim, S., Kim, S., 2010. The follow-up study on patients of amyotrophic lateral sclerosis after 1 year. J. Pharmacopuncture 13, 121–128.

Kim, S., Chung, S.E., Lee, S., Park, J., Choi, S., Kim, S., 2016. Experience of complementary and alternative medicine in patients with amyotrophic lateral sclerosis and their families: a qualitative study. Amyotrophic Lateral Scler. Frontotemporal Degener. 2016 (17), 191–197.

Kok, J.R., Palminha, N.M., Santos Souza, C.D., El-Khamisy, S.F., Ferraiuolo, L., 2021. DNA damage as a mechanism of neurodegeneration in ALS and a contributor to astrocyte toxicity. Cell. Mol. Life Sci. 78, 5707–5729.

Kook, M.G., Choi, S.W., Seo, Y., Kim, D.W., Song, B.K., Son, I., et al., 2017. KCHO-1, a novel herbal anti-inflammatory compound, attenuates oxidative stress in an animal model of amyotrophic lateral sclerosis. J. Vet. Sci. 18, 487–497.

Kuboyama, T., Tohda, C., Komatsu, K., 2014. Effects of Ashwagandha (roots of Withania somnifera) on neurodegenerative diseases. Biol. Pharm. Bull. 37, 892–897.

Lahlou, M., 2013. The success of natural products in drug discovery. Pharmacol. Pharm. 4, 17–31.

Lee, D.S., Ko, W., Yoon, C.S., Kim, D.C., Yun, J., Lee, J.K., et al., 2014. KCHO-1, a novel antineuroinflammatory agent, inhibits lipopolysaccharide-induced neuroinflammatory responses through Nrf2-mediated heme oxygenase-1 expression in mouse BV2 microglia cells. Evid. Based Complement Alternat. Med., 357154.

Lee, S.J., Jeong, H.H., Lee, J.C., Cha, E.H., Park, M.Y., Song, B.G., et al., 2016a. A study on single dose toxicity of intravenous injection of mecasin herbal acupuncture. J. Korean Acupunct. Moxibustion Med. 33, 1–7.

Lee, D.S., Ko, W., Song, B.K., Son, I., Kim, D.W., Kang, D.G., et al., 2016b. The herbal extract KCHO-1 exerts a neuroprotective effect by ameliorating oxidative stress via heme oxygenase-1 upregulation. Mol. Med. Rep. 13, 4911–4919.

Lopez-Gonzalez, R., Lu, Y., Gendron, T.F., Karydas, A., Tran, H., Yang, D., et al., 2016. Poly(GR) in C9ORF72-related ALS/FTD compromises mitochondrial function and increases oxidative stress and DNA damage in iPSC derived motor neurons. Neuron 92, 383–391.

Mancuso, R., del Valle, J., Modol, L., Martinez, A., Granado-Serrano, A.B., Ramirez-Núñez, O., et al., 2014a. Resveratrol improves motoneuron function and extends survival in SOD1G93A ALS mice. Neurotherapeutics 11, 419–432.

Maor-Nof, M., Shipony, Z., Lopez-Gonzalez, R., Nakayama, L., Zhang, Y., Couthouis, J., et al., 2021. p53 is a central regulator driving neurodegeneration caused by C9orf72 poly(PR). Cell 184, 689–708.

Maragakis, N.J., Rothstein, J.D., 2006. Mechanisms of disease: astrocytes in neurodegenerative disease. Nat. Clin. Pract. Neurol. 2, 679–689.

Markert, C.D., Kim, E., Gifondorwa, D.J., Childers, M.K., Milligan, C.E., 2010. A single-dose resveratrol treatment in a mouse model of amyotrophic lateral sclerosis. J. Med. Food 13, 1081–1085.

Martinez, H.R., Escamilla-Ocanas, C.E., Tenorio-Pedraza, J.M., Gomez-Almaguer, D., Jaime-Perez, J.C., Olguin-Ramirez, et al., 2016. Altered CSF cytokine network in amyotrophic lateral sclerosis patients: a pathway-based statistical analysis. Cytokine 90, 1–5.

Marton, R.M., Pasca, S.P., 2016. Neural differentiation in the third dimension: generating a human midbrain. Cell Stem Cell 19, 145–146.

Mathur, S., Hoskins, C., 2017. Drug development: lessons from nature. Biomed. Rep. 6, 612–614.

Miller, R.G., Mitchell, J.D., Moore, D.H., 2002. Riluzole for amyotrophic lateral sclerosis (ALS)/motor neuron disease (MND). Cochrane Database Syst. Rev., CD001447.

Mishra, L.C., Singh, B.B., Dagenais, S., 2000. Scientific basis for the therapeutic use of Withania somnifera (ashwagandha): a review. Altern. Med. Rev. 5, 334–346.

Mitsuyama, Y., 1993. Presenile dementia with motor neuron disease. Dement. Geriatr. Cogn. Disord. 4, 137–142.

Muthaiyah, B., Essa, M.M., Chauhan, V., Chauhan, A., 2011. Protective effects of walnut extract against amyloid beta peptide-induced cell death and oxidative stress in PC12 cells. Neurochem. Res. 36, 2096–2103.

Neary, D., Snowden, J.S., Mann, D.M.A., Northern, B., Goulding, P.J., Macdermott, N., 1990. Frontal lobe dementia and motor neuron disease. J. Neurol. Neurosurg. Psychiatry 53, 23–32.

Newman, D.J., Cragg, G.M., 2016. Natural products as sources of new drugs from 1981 to 2014. J. Nat. Prod. 79, 629–661.

Nihei, Y., Mori, K., Werner, G., Arzberger, T., Zhou, Q., Khosravi, B., et al., 2020. Poly-glycine–alanine exacerbates C9orf72 repeat expansion-mediated DNA damage via sequestration of phosphorylated ATM and loss of nuclear hnRNPA3. Acta Neuropathol. 139, 99–118.

Norden, D.M., Muccigrosso, M.M., Godbout, J.P., 2015. Microglial priming and enhanced reactivity to secondary insult in aging, and traumatic CNS injury, and neurodegenerative disease. Neuropharmacology 96, 29–41.

Novak, V., Rogelj, B., Zupunski, V., 2021. Therapeutic potential of polyphenols in amyotrophic lateral sclerosis and frontotemporal dementia. Antioxidants 10, 1328.

Obulesu, M., 2021. Turmeric and Curcumin in the Treatment of Neurodegenerative Diseases, first ed. Elsevier.

Ono, S., Hu, J., Shimizu, N., Imai, T., Nakagawa, H., 2001. Increased interleukin-6 of skin and serum in amyotrophic lateral sclerosis. J. Neurol. Sci. 187, 27–34.

Pehar, M., Vargas, M.R., Cassina, P., Barbeito, A.G., Beckman, J.S., Barbeito, L., 2005. Complexity of astrocyte-motor neuron interactions in amyotrophic lateral sclerosis. Neurodegener. Dis. 2, 139–146.

Perry, V.H., Holmes, C., 2014. Microglial priming in neurodegenerative disease. Nat. Rev. Neurol. 10, 217–224.

Petrov, D., Mansfield, C., Moussy, A., Hermine, O., 2017. ALS clinical trials review: 20 years of failure. Are we any closer to registering a new treatment? Front. Aging Neurosci. 9, 1–11.

Philips, T., Rothstein, J.D., 2014. Glial cells in amyotrophic lateral sclerosis. Exp. Neurol. 262, 111–120.

Pohl, F., Thoo Lin, P.K., 2018. The potential use of plant natural products and plant extracts with antioxidant properties for the prevention/treatment of neurodegenerative diseases: in vitro, in vivo and clinical trials. Molecules 23, 3283.

Rambaran, T.F., 2020. Nanopolyphenols: a review of their encapsulation and anti-diabetic effects. SN Appl. Sci. 2, 1–26.

Renton, A.E., Majounie, E., Waite, A., Simon-Sanchez, J., Rollinson, S., Gibbs, J.R., et al., 2011. A hexanucleotide repeat expansion in C9ORF72 is the cause of chromosome 9p21-linked ALS-FTD. Neuron 72, 257–268.

Rentzos, M., Nikolaou, C., Rombos, A., Boufidou, F., Zoga, M., Dimitrakopoulos, A., et al., 2007. RANTES levels are elevated in serum and cerebrospinal fluid in patients with amyotrophic lateral sclerosis. Amyotroph. Lateral Scler. 8, 283–287.

Rentzos, M., Rombos, A., Nikolaou, C., Zoga, M., Zouvelou, V., Dimitrakopoulos, A., et al., 2010. Interleukin-17 and interleukin-23 are elevated in serum and cerebrospinal fluid of patients with ALS: a reflection of Th17 cells activation? Acta Neurol. Scand. 122, 425–429.

Rosen, D.R., Siddique, T., Patterson, D., Figlewicz, D.A., Sapp, P., Hentati, A., et al., 1993. Mutations in Cu/Zn superoxide dismutase gene are associated with familial amyotrophic lateral sclerosis. Nature 362, 59–62.

Saeidnia, S., Gohari, A.R., 2012. Importance of Brassica napus as a medicinal food plant. J. Med. Plants Res. 6, 2700–2703.

Schiaffino, L., Bonafede, R., Scambi, I., Parrella, E., Pizzi, M., Mariotti, R., 2018. Acetylation state of RelA modulated by epigenetic drugs prolongs survival and induces a neuroprotective effect on ALS murine model. Sci. Rep. 8, 1–13.

Silva, R.F.M., Pogacnik, L., 2020. Polyphenols from food and natural products: neuroprotection and safety. Antioxidants 9, 61.

Singh, N., Bhalla, M., de Jager, P., Gilca, M., 2011. An overview on ashwagandha: a Rasayana (rejuvenator) of Ayurveda. Afr. J. Tradit. Complement. Altern. Med. 8, 208–213.

Solanki, I., Parihar, P., Mansuri, M.L., Parihar, M.S., 2015. Flavonoid-based therapies in the early management of neurodegenerative diseases. Adv. Nutr. 6, 64–72.

Soo, S.K., Rudich, P.D., Traa, A., Harris-Gauthier, N., Shields, H.J., Van Raamsdonk, J.M., 2020. Compounds that extend longevity are protective in neurodegenerative diseases and provide a novel treatment strategy for these devastating disorders. Mech. Ageing Dev. 190, 111297.

Sordet, O., Redon, C.E., Guirouilh-Barbat, J., Smith, S., Solier, S., Douarre, C., et al., 2009. Ataxia telangiectasia mutated activation by transcription- and topoisomerase I-induced DNA double-strand breaks. EMBO Rep. 10, 887–893.

Stephens, H.E., Young, J., Felgoise, S.H., Simmons, Z., 2016. A qualitative study of multidisciplinary ALS clinic use in the United States. Amyotroph. Lateral Scler. Frontotemporal Degener. 17, 55–61.

Sun-Waterhouse, D., 2011. The development of fruit-based functional foods targeting the health and wellness market: a review. Int. J. Food Sci. Technol. 46, 899–920.

Swallah, M.S., Sun, H., Affoh, R., Fu, H., Yu, H., 2020. Antioxidant potential overviews of secondary metabolites (polyphenols) in fruits. Int. J. Food Sci. 2020, 1–8.

Swarup, V., Phaneuf, D., Bareil, C., Robertson, J., Rouleau, G.A., Kriz, J., et al., 2011. Pathological hallmarks of amyotrophic lateral sclerosis/frontotemporal lobar degeneration in transgenic mice produced with TDP-43 genomic fragments. Brain 134, 2610–2626.

Tandan, R., Bradley, W.G., 1985. Amyotrophic lateral sclerosis: part 1. Clinical features, pathology, and ethical issues in management. Ann. Neurol. 18, 271–280.

Teng, Y., Yadav, T., Duan, M., Tan, J., Xiang, Y., Gao, B., et al., 2018. ROS-induced R loops trigger a transcription-coupled but BRCA1/2-independent homologous recombination pathway through CSB. Nat. Commun. 9, 1–12.

Tsao, R., 2010. Chemistry and biochemistry of dietary polyphenols. Nutrients 2, 1231–1246.

Upadhyay, S., Dixit, M., 2015. Role of polyphenols and other phytochemicals on molecular signaling. Oxidative Med. Cell. Longev. 2015, 504253.

Upton, R., Graff, A., Evans, F., et al., 2000. In: Upton, R. (Ed.), Ashwagandha root-Withania somnifera: Analytical, Quality Control and Therapeutic Monograph in American Herbal Pharmacopoeia and Therapeutic Compendium. AHP, pp. 1–26. Epub 2000.

Usdin, K., House, N.C.M., Freudenreich, C.H., 2015. Repeat instability during DNA repair: insights from model systems. Crit. Rev. Biochem. Mol. Biol. 50, 142–167.

Walker, C., El-Khamisy, S.F., 2018. Perturbed autophagy and DNA repair converge to promote neurodegeneration in amyotrophic lateral sclerosis and dementia. Brain 141, 1247–1262.

Walker, C., Herranz-Martin, S., Karyka, E., Liao, C., Lewis, K., Elsayed, W., et al., 2017. C9orf72 expansion disrupts ATM-mediated chromosomal break repair. Nat. Neurosci. 20, 1225–1235.

Wang, J., Zhang, Y., Tang, L., Zhang, N., Fan, D., 2011. Protective effects of resveratrol through the up-regulation of SIRT1 expression in the mutant hSOD1-G93A-bearing motor neuron-like cell culture model of amyotrophic lateral sclerosis. Neurosci. Lett. 503, 250–255.

Wink, M., Abbas, S., 2013. Epigallocatechin gallate (EGCG) from green tea (Camellia sinensis) and other natural products mediate stress resistance and slows down aging processes

in caenorhabditis elegans. In: Preedy, V.R. (Ed.), Tea in Health and Disease Prevention. Elsevier Science Publishing Co., Inc, London, UK, pp. 1105–1115.
Yáñez, M., Galán, L., Matías-Guiu, J., Vela, A., Guerrero, A., García, A.G., 2011. CSF from amyotrophic lateral sclerosis patients produces glutamate independent death of rat motor brain cortical neurons: protection by resveratrol but not riluzole. Brain Res. 1423, 77–86.
Yang, X., Gao, L., Wu, X., Zhang, Y., Zang, D., 2016. Increased levels of MIP-1 alpha in CSF and serum of ALS. Acta Neurol. Scand. 134, 94–100.
Yang, B., Dong, Y., Wang, F., Zhang, Y., 2020. Nanoformulations to enhance the bioavailability and physiological functions of polyphenols. Molecules 25, 4613.
Yoshino, H., Kimura, A., 2006. Investigation of the therapeutic effects of edaravone, a free radical scavenger, on amyotrophic lateral sclerosis (Phase II study). Amyotroph. Lateral Scler. 7, 247–251.
Yue, J., Shang, P., Wang, G., Liu, D., Xu, L., Zhou, W., 2014. Nutritional and antioxidant properties of rapeseed (Brassica Napus) cultivars with high and low erucic acid content. J. Food Nutr. Res. 2, 918–924.
Zarei, S., Carr, K., Reiley, L., Diaz, K., Guerra, O., Altamirano, P.F., et al., 2015. A comprehensive review of amyotrophic lateral sclerosis. Surg. Neurol. Int. 6, 171.

Further reading

Mancuso, R., Del Valle, J., Morell, M., Pallás, M., Osta, R., Navarro, X., 2014b. Lack of synergistic effect of resveratrol and sigma-1 receptor agonist (PRE-084) in SOD1G93A ALS mice: overlapping effects or limited therapeutic opportunity? Orphanet J. Rare Dis. 9, 1–11.

CHAPTER 9

Neuroprotective role of curcumin in amyotrophic lateral sclerosis

Introduction

Pathology

Amyotrophic lateral sclerosis (ALS), also known as Lou Gehrig's disease or Charcot disease, is a lethal neurodegenerative disease (ND) that perturbs motor neurons (MNs) and results in death within 2–5 years of diagnosis due to the lack of substantial therapy (Cunha-Oliveira et al., 2020). While ALS patients survive only 2–3 years from the time of diagnosis, 5%–10% of patients survive beyond 10 years after diagnosis. This incurable ND shows a worldwide incidence of four to six cases in 100,000 and naturally has late onset with symptoms observed between 55 and 65 years of age (Smith et al., 2019). Usually, men undergo an earlier age of onset unlike women, and they are more likely to experience spinal onset, while bulbar onset is more frequent in women (McCombe and Henderson, 2010). Although the exceedingly common form of ALS is sporadic (sALS), with unknown etiology in approximately 90%–95% of cases, the residual 5%–10% of cases are inherited (familial ALS—fALS) and normally linked to an earlier age of onset (Zarei et al., 2015; Valdmanis and Rouleau, 2008). Despite scant evidence on the causes of sALS, a few risk factors such as age, smoking, body mass index, state of physical fitness, and occupational and environmental exposure to chemicals, pesticides, metals, and electromagnetic fields were found to contribute to the etiology of the disease (Ingre et al., 2015). Nevertheless, these external parameters do not play a direct role in provoking the onset of this disease; it is supposed that there are a few distinct susceptibility factors associated with external exposure to environmental factors resulting in the development of ALS (Al-Chalabi and Hardiman, 2013; Lacorte et al., 2016; Yu and Pamphlett, 2017).

ALS is an intensifying ND triggered by the deprivation of upper MNs in the cerebral cortex and lower MNs in the brain stem and spinal cord, resulting in muscle feebleness and eventually inducing muscle atrophy and paralysis that ends in respiratory failure and death (Boillee et al., 2006; Tafuri

https://doi.org/10.1016/B978-0-323-95762-5.00009-6

et al., 2015; Zarei et al., 2015; Cunha-Oliveira et al., 2020). The most commonly noticed characteristic of the disease in sporadic and familial ALS patients is the robust activation and proliferation of microglia at the regions of MN loss (Henkel et al., 2004). Microglial activation may be provoked by the deposition of aggregates of mutant superoxide dismutase (SOD1) (Heneka et al., 2014). These activated microglia exhibit neuroprotective efficacy during the early presymptomatic stages of ALS but change to a more proinflammatory phenotype as the disease continues, provoking neurodegeneration (Liao et al., 2012; Haukedal and Freude, 2019; Zhao et al., 2020). Several molecular mechanisms have been recommended such as glutamate excitotoxicity, changed RNA metabolism, impaired axonal transport, protein misfolding and aggregation, endoplasmic reticulum stress, disordered protein trafficking, oxidative stress (OS), inflammation, and mitochondrial impairment (Smith et al., 2019; Lu et al., 2016; Browne and Abbott, 2016).

Oxidative stress

A wealth of studies have shown enhanced levels of oxidative damage in proteins, lipids, and DNA of postmortem neuronal tissue (Ferrante et al., 1997; Shaw et al., 1995; Calingasan et al., 2005) and in cerebrospinal fluid (CSF) (Smith et al., 1998; Tohgi et al., 1999; Simpson et al., 2004; Ihara et al., 2005), plasma (Weishaupt et al., 2006), and urine (Mitsumoto et al., 2009) samples received from ALS patients, indicating the role of OS mechanisms in the central nervous system (CNS) and other tissues. Nevertheless, it is a herculean task to examine if oxidative damage has a primary role or secondary importance in the pathology of this disease (Agar and Durham, 2009; Carrì et al., 2003) and if oxidative damage occurs early or late in the course of the disease.

MutSOD1, contributing to around 20% of ALS cases, is one of the widely studied causes of ALS, including OS mechanisms and interruption of mitochondrial function noticed in cultured cells (Chi et al., 2007; Ferri et al., 2006; Pesaresi et al., 2011) and in animal models (Igoudjil et al., 2011; Vargas et al., 2011; Pickles et al., 2013).

Therapy

Since OS plays a pivotal role in ALS, antioxidant-based therapeutics can substantially combat the disease (Cunha-Oliveira et al., 2020). Targeting microglial activation by inhibiting proinflammatory neurotoxic effects of the typically initiated phenotype and/or concurrently increasing the antiinflammatory neuroprotective functions of the interchangeably activated phenotype are vital therapeutic avenues for NDs (Daneman and Prat, 2015).

Curcumin

A few compounds such as resveratrol, curcumin, and ferulic acid have been found to scavenge free radicals through electron transfer and guard cellar components like DNA, RNA, and cell membranes from ROS/RNS-instigated oxidative damage (Lu et al., 2010; Fu et al., 2018; Zhao et al., 2020).

Curcumin exerts a plethora of vital pharmacological activities such as anticancer (Moustapha et al., 2015; Rainey et al., 2015; Shanmugam et al., 2015; Ahmadi et al., 2017), antimicrobial (Zorofchian Moghadamtousi et al., 2014), antiinflammatory (Mukhopadhyay et al., 1982), antiamyloid (Yang et al., 2005; Aggarwal et al., 2007), antioxidant (Menon and Sudheer, 2007), and neuroprotective effects (Thiyagarajan and Sharma, 2004). Therefore, it has been extensively used in the treatment of NDs such as Alzheimer's disease (AD), Parkinson's disease (PD), and Huntington's disease (HD) (Joe et al., 2004; Jaroonwitchawan et al., 2017). It has long been used for numerous aliments, and it is extensively taken for dietary and medicinal purposes in Southeast Asia, China, and India (Anand et al., 2007; Wang et al., 2017; Yavarpour-Bali et al., 2019). It has the efficacy to span the blood-brain barrier (BBB), and because of its multiple therapeutic effects, curcumin has been considered as a substantial therapeutic aspect for manifold nervous system diseases (Maiti and Dunbar, 2018). Moreover, curcumin has incredible use in the treatment of numerous diseases like diabetes, cancer, malaria, cystic fibrosis, and hypertension (Yallapu et al., 2013; Shehzad et al., 2013).

Curcumin, a commonly used liposoluble dye derived from turmeric, is another compound that controls the Nrf2 pathway (Carvalho et al., 2017). Curcumin was found to initiate the Nrf2 pathway in primary spinal cord astrocytes, weakening oxidative damage and mitochondrial impairment (Jiang et al., 2011). Besides these advantages, curcumin was also found to bind to the prefibrillar aggregates of SOD1 and modify their amyloidogenic pathway, reducing cytotoxicity (Bhatia et al., 2015).

Dimethoxy curcumin

Dimethoxy curcumin ameliorated mitochondrial impairment in NSC-34 cell line transfected with human M337V or Q331K mutant TDP-43, indicating that this compound can be used in the treatment of ND associated with mutated TDP-43 (Lu et al., 2012). The oral introduction of 80 mg/day nanocurcumin (SinaCurcumin) in a pilot randomized clinical trial on 54 sALS patients for 12 months demonstrated an amelioration in the survival of ALS patients, particularly those with bulbar involvement (https://en.irct.ir/trial/11697) (Ahmadi et al., 2018).

In addition, in a double-blind clinical trial, curcumin oral administration (600 mg/day, Brainoil) in 42 ALS patients for 6 months led to a reduction in ALS progression, amelioration of aerobic metabolism, and a decrease of oxidative damage (Chico et al., 2018). Although curcumin has therapeutic benefits, unstable chemical structure, low oral bioavailability, and poor water solubility are the shortcomings that require drug delivery systems to enhance therapeutic efficacy (Rakotoarisoa and Angelova, 2018; Tripodo et al., 2015). Curcuminoids in addition to the crude extracts of *C. longa* have been widely studied for their neuroprotective effects (Ojha et al., 2012; Pan et al., 2012; Tripanichkul and Jaroensuppaperch, 2012; Elufioye et al., 2017).

Curcumin against TDP-43

It exerts neuroprotective effects and combats OS (Kim et al., 2012), mitochondrial dysfunction, inflammation, and protein aggregation (Ghasemi et al., 2019; Rakotoarisoa and Angelova, 2018; Ullah et al., 2017). It eradicates the excitability instigated by TDP-43 (the vital pathological protein in sporadic ALS) in an MN-like cellular model of ALS (Dong et al., 2014; Mackenzie and Rademakers, 2008). In MN cell lines transfected with the TDP-43 mutant, curcumin decreases OS and rescues cells from mitochondrial dysfunction (Lu et al., 2012; Duan et al., 2014). Curcumin may ameliorate life expectancy in patients with ALS, specifically those with bulbar involvement (Ahmadi et al., 2018). In a double-blind therapeutic trial, curcumin treatment demonstrated a reduction in ALS progression and a decrease of oxidative damage (Chico et al., 2018).

A study suggests that a curcumin-based drug delivery system can be recommended for the treatment of ALS (Tripodo et al., 2015), but others showed that the use of curcumin as therapy in ALS has drawbacks, such as chemical instability and low oral bioavailability and water solubility rate (Rakotoarisoa and Angelova, 2018). Consequently, there is a growing need to discover novel technologies to circumvent these limitations (Liu et al., 2016).

Recent studies used curcumin in the treatment of NDs. Although pathological events vary for each disease, a few shared events facilitate curcumin to induce neuroprotective effects. In line with this, inflammation is a shared characteristic in numerous NDs. Most commonly, NFκB induces the inflammatory response by initiating TNF-α, which further regulates ROS activation. It shows an attenuating effect on NFκB, thus curtailing inflammation (Luss et al., 2018). Neurological diseases, like multiple sclerosis (MS), spinal cord injury (SCI), and AD (Dolati et al., 2018), demonstrate that an immune cell attack and additional mechanism can be controlled by curcumin via T-lymphocyte activation and monocyte chemotaxis. Curcumin shows negative control of pathways that regulate chemokines, thus playing a role in monocyte employment and activation of T-lymphocytes (Bianconi et al., 2018; Balasubramanian and Eckert, 2007; Gagliardi et al., 2020).

Use of adjuvants

The utilization of the antioxidant nanocurcumin (or vitamin E) as a supplement therapy to riluzole in ALS patients failed to ameliorate motor function (Graf et al., 2005; Ahmadi et al., 2018; Jagaraj et al., 2020). Currently, a plethora of strategies such as employment of adjuvant molecules like quercetin, piperine, or silibinin; use of curcumin's structural analogues or its complexation with phospholipids, polysaccharides, or proteins; and bioconjugates of curcumin with alanine or turmeric oil have been used to enhance solubility and bioavailability of curcumin (Naksuriya et al., 2014; Yallapu et al., 2015). Despite the increase in solubility and bioavailability of curcumin by these aforementioned strategies, there was a lack of absolute protection for curcumin from the swift metabolism in the body. Additionally, these approaches failed to funnel curcumin to the target site.

Nanotechnology

To overcome the limitation of curcumin and target curcumin to the mitochondrial matrix, use of biodegradable, polymeric nanoparticles (NPs) and surface functionalization with the lipophilic triphenylphosphonium cation have been studied (Marrache and Dhar, 2012; Wilkins and Morris, 2017). Drug delivery systems (DDS) designed using NPs have been used to enhance water solubility and bioavailability of therapeutic molecules like curcumin. Mounting evidence has shown that incorporation of curcumin into NPs profoundly ameliorates chemical stability of curcumin and averts its enzymatic and pH degradation. Moreover, curcumin-loaded NPs enhance its circulation inside the body (Anand et al., 2007; Szymusiak et al., 2016; Maiti and Dunbar, 2018).

A plethora of nanoformulations of curcumin have been employed to enhance the aqueous solubility and brain levels of curcumin to treat NDs. In line with this, previous studies have shown a profound improvement in the bioavailability of orally introduced curcumin in plasma (Xie et al., 2011; Shaikh et al., 2009) and bioavailability of intraperitoneally or intravenously administered curcumin in the brain when it was incorporated in poly(lactide-*co*-glycolide) (PLGA) NPs when compared with curcumin alone (Tsai et al., 2011; Tiwari et al., 2014). Since PLGA is cleaved into glycolic acid and lactic acid and are uptaken into TCA cycle for eventual excretion as H_2O and CO_2, they are regarded as safe. The addition of groups like polyethylene glycol (PEG) rescues PLGA NPs against early degradation through the reticuloendothelial system (Essa et al., 2020) and enhances the plasma bioavailability of curcumin even more after oral administration (Khalil et al., 2013). Encapsulation of curcumin into human serum albumin NPs augmented plasma and peripheral tissue levels after intravenous injection (Kim et al., 2011), thereby corroborating that serum albumin is internalized through endothelia (Vogel et al., 2001). Pellets of solid dispersion of curcumin using polyvinylpyrrolidone-loaded pellets of solid curcumin also improved plasma levels when compared to conventional curcumin after oral administration (Liao et al., 2019).

Nanocurcumin in the treatment of ALS

Mesenchymal stromal cells (MSCs) showed a novel avenue in amelioration of CNS repair. MSCs travel to the injured tissues and effectively span the BBB (Chapel et al., 2003; de Girolamo et al., 2013). The pivotal role of MSCs in the amelioration of CNS diseases like ALS was also exhibited by increasing neural protection and substituting dead MNs of the spinal cord (Tripodo et al., 2015). In line with this, insulin-D-alpha-tocopherol succinate micelles (INVITE)-loaded curcumin micelles have been investigated to enhance the therapeutic efficacy of MSCs. This strategy can assure appropriate drug targeting regulated by the characteristic potency of MSCs to arrive at damaged tissues while defending the encapsulated drug from untimely release or avoiding toxicity of MSCs. This novel DDS has been considered as an exemplary armamentarium for combating NDs like ALS (Tripodo et al., 2015; Yavarpour-Bali et al., 2019).

DMC protects against damage initiated by the overexpression of TDP-43 by ameliorating mitochondrial membrane potential and reducing the levels of mitochondrial uncoupling protein 2 (UCP2) (Lu et al., 2012) that exerts neuroprotective efficacy by controlling mitochondrial transmembrane potential, mitochondrial proliferation, and ATP and ROS levels (Conti et al., 2005). Curcumin also plays an essential role in the initiation and propagation of action potentials that are augmented by the overexpression of wild-type and mutant Q331K TDP-43; certainly, irregular action potentials and voltage-gated sodium channels on motoneuron-like cell lines was remarkably improved after DMC treatment, which probably functions by mitigating OS and mitochondrial impairment (Dong et al., 2014). A randomized pilot clinical trial conducted recently exhibited the safety and efficacy of nanocurcumin in ameliorating survival if used as add-on in ALS patients with bulbar symptoms.

Conclusions and future perspectives

Despite the lack of adequate studies on pathological mechanisms of ALS, several lines of evidence showed that ROS generation linked to inadequate antioxidant activity has been considered a vital pathological event in ALS. Many antioxidant agents could not exert sufficient therapeutic benefit or be adequately examined. The difficulty in analyzing the OS markers at an early stage of the disease has been a major hurdle in the diagnosis and treatment of ALS (Cunha-Oliveira et al., 2020). Combination therapy using curcumin-loaded NPs is suitable for reducing the dose of the primary therapeutic that can lead to ameliorated therapeutic ability as well as decreasing systemic toxicity. In addition despite the success of functionalized NPs in drug targeting, their nanosize structure and the huge surface area probably result in particle accumulation and lower drug loading (Roney et al., 2005; Kaur et al., 2008; Yavarpour-Bali et al., 2019).

Although curcumin exhibits noteworthy pharmacological efficacy, low water solubility, instability in the body fluids, enhanced metabolism, swift clearance, decreased absorption in gastrointestinal (GI) tract, and reduced bioavailability impeded its success (Anand et al., 2007; Maiti and Dunbar, 2018). Despite the confirmation of curcumin's safety by many studies, these biological riddles prevented its escalation to clinical use (Dende et al., 2017). Therefore, robust curcumin delivery systems are currently significant.

References

Agar, J., Durham, H., 2009. Relevance of oxidative injury in the pathogenesis of motor neuron diseases. Amyotroph. Lateral Scler. Other Motor Neuron Disord. 4 (4), 232–242.

Aggarwal, B.B., Sundaram, C., Malani, N., Ichikawa, H., 2007. Curcumin: the Indian solid gold. In: The Molecular Targets and Therapeutic Uses of Curcumin in Health and Disease. vol. 595. Springer, pp. 1–75.

Ahmadi, F., Ghasemi-Kasman, M., Ghasemi, S., Tabari, M.J., Pourbagher, R., Kazemi, S., et al., 2017. Induction of apoptosis in Hela cancer cells by an ultrasonic-mediated synthesis of curcumin-loaded chitosan–alginate–sTPP nanoparticles. Int. J. Nanomedicine 12, 8545.

Ahmadi, M., Agah, E., Nafissi, S., Jaafari, M.R., Harirchian, M.H., Sarraf, P., et al., 2018. Safety and efficacy of nanocurcumin as add-on therapy to riluzole in patients with amyotrophic lateral sclerosis: a pilot randomized clinical trial. Neurotherapeutics 15, 430–438.

Al-Chalabi, A., Hardiman, O., 2013. The epidemiology of ALS: a conspiracy of genes, environment and time. Nat. Rev. Neurol. 9, 617–628.

Anand, P., Kunnumakkara, A.B., Newman, R.A., Aggarwal, B.B.J., 2007. Bioavailability of curcumin: problems and promises. Mol. Pharm. 4, 807–818.

Balasubramanian, S., Eckert, R.L., 2007. Keratinocyte proliferation, differentiation, and apoptosis—differential mechanisms of regulation by curcumin, EGCG and apigenin. Toxicol. Appl. Pharmacol. 224, 214–219.

Bhatia, N.K., Srivastava, A., Katyal, N., Jain, N., Khan, M.A., Kundu, B., et al., 2015. Curcumin binds to the pre-fibrillar aggregates of Cu/Zn superoxide dismutase (SOD1) and alters its amyloidogenic pathway resulting in reduced cytotoxicity. Biochim. Biophys. Acta 1854, 426–436.

Bianconi, V., Sahebkar, A., Atkin, S.L., Pirro, M., 2018. The regulation and importance of monocyte chemoattractant protein-1. Curr. Opin. Hematol. 25, 44–51.

Boillee, S., Yamanaka, K., Lobsiger, C.S., Copeland, N.G., Jenkins, N.A., Kassiotis, G., et al., 2006. Onset and progression in inherited ALS determined by motor neurons and microglia. Science 312, 1389–1392.

Browne, E.C., Abbott, B.M., 2016. Recent progress towards an effective treatment of amyotrophic lateral sclerosis using the SOD1 mouse model in a preclinical setting. Eur. J. Med. Chem. 121, 918–925.

Calingasan, N.Y., Chen, J., Kiaei, M., Beal, M.F., 2005. Beta amyloid 42 accumulation in the lumbar spinal cord motor neurons of amyotrophic lateral sclerosis patients. Neurobiol. Dis. 19, 340–347.

Carrì, M.T., Ferri, A., Cozzolino, M., Calabrese, L., Rotilio, G., 2003. Neurodegeneration in amyotrophic lateral sclerosis: the role of oxidative stress and altered homeostasis of metals. Brain Res. Bull. 61, 365–374.

Carvalho, A.N., Firuzi, O., Gama, M.J., Horssen, J.V., Saso, L., 2017. Oxidative stress and antioxidants in neurological diseases: is there still hope? Curr. Drug Targets 18, 705–718.

Chapel, A., Bertho, J.M., Bensidhoum, M., Fouillard, L., Young, R.G., Frick, J., et al., 2003. Mesenchymal stem cells home to injured tissues when co-infused with hematopoietic cells to treat a radiation-induced multi-organ failure syndrome. J. Gene Med. 5, 1028–1038. A cross disciplinary journal for research on the science of gene transfer and its clinical applications.

Chi, L., Ke, Y., Luo, C., Gozal, D., Liu, R., 2007. Depletion of reduced glutathione enhances motor neuron degeneration in vitro and in vivo. Neuroscience 144, 991–1003.

Chico, L., Lenco, E.C., Bisordi, C., Lo Gerfo, A., Petrozzi, L., Petrucci, A., et al., 2018. Amyotrophic lateral sclerosis and oxidative stress: a double-blind therapeutic trial after curcumin supplementation. CNS Neurol. Disord. Drug Targets 17, 767–779.

Conti, B., Sugama, S., Lucero, J., Winsky-Sommerer, R., Wirz, S.A., Maher, P., et al., 2005. Uncoupling protein 2 protects dopaminergic neurons from acute 1,2,3,6-methyl-phenyl-tetrahydropyridine toxicity. J. Neurochem. 93, 493–501.

Cunha-Oliveira, T., Montezinho, L., Mendes, C., Firuzi, O., Saso, L., Oliveira, P.J., et al., 2020. Oxidative stress in amyotrophic lateral sclerosis: pathophysiology and opportunities for pharmacological intervention. Oxid. Med. Cell Longev. 2020, 5021694.

Daneman, R., Prat, A., 2015. The blood–brain barrier. Cold Spring Harb. Perspect. Biol. 7, a020412.

de Girolamo, L., Lucarelli, E., Alessandri, G., Avanzini, M.A., Bernardo, M.E., Biagi, E., et al., 2013. Mesenchymal stem/stromal cells: a new "cells as drugs" paradigm. Efficacy Crit. Aspects Cell Ther. 19, 2459–2473.

Dende, C., Meena, J., Nagarajan, P., Nagaraj, V.A., Panda, A.K., Padmanaban, G.J., 2017. Nanocurcumin is superior to native curcumin in preventing degenerative changes in experimental cerebral malaria. Sci. Rep. 7, 10062.

Dolati, S., Marofi, F., Babaloo, Z., Aghebati-Maleki, L., Roshangar, L., Ahmadi, M., et al., 2018. Dysregulated network of miRNAs involved in the pathogenesis of multiple sclerosis. Biomed. Pharmacother. 104, 280–290.

Dong, H., Xu, L., Wu, L., Wang, X., Duan, W., Li, H., et al., 2014. Curcumin abolishes mutant TDP-43 induced excitability in a motoneuron-like cellular model of ALS. Neuroscience 272, 141–153.

Duan, W., Guo, Y., Xiao, J., Chen, X., Li, Z., Han, H., et al., 2014. Neuroprotection by monocarbonyl dimethoxycurcumin C: ameliorating the toxicity of mutant TDP-43 via HO-1. Mol. Neurobiol. 49, 368–379.

Elufioye, T.O., Berida, T.I., Habtemariam, S., 2017. Plants-derived neuroprotective agents: cutting the cycle of cell death through multiple mechanisms. Evid. Based Complement. Alternat. Med. 2017, 3574012.

Essa, D., Kondiah, P.P.D., Choonara, Y.E., Pillay, V., 2020. The design of poly(lactide-co-glycolide) nanocarriers for medical applications. Front. Bioeng. Biotechnol. 8, 48.

Ferrante, R.J., Browne, S.E., Shinobu, L.A., Bowling, A.C., Baik, M.J., MacGarvy, U., et al., 1997. Evidence of increased oxidative damage in both sporadic and familial amyotrophic lateral sclerosis. J. Neurochem. 69, 2064–2074.

Ferri, A., Cozzolino, M., Crosio, C., Nencini, M., Casciati, A., Gralla, E.B., et al., 2006. Familial ALS superoxide dismutases associate with mitochondria and shift their redox potentials. Proc. Natl Acad. Sci. U. S. A 103, 13860–13865.

Fu, Y., Yang, J.M., Wang, X.Y., Yang, P., Zhao, Y., Li, K., et al., 2018. Herbal compounds play a role in neuroprotection through the inhibition of microglial activation. J. Immunol. Res. 2018, 9308046.

Gagliardi, S., Morasso, C., Stivaktakis, P., Pandini, C., Tinelli, V., Tsatsakis, A., et al., 2020. Curcumin formulations and trials: what's new in neurological diseases. Molecules 25, 5389.

Ghasemi, F., Bagheri, H., Barreto, G.E., Read, M.I., Sahebkar, A.J.N., 2019. Effects of curcumin on microglial cells. Neurotox. Res. 36, 12–26.

Graf, M., Ecker, D., Horowski, R., Kramer, B., Riederer, P., Gerlach, M., et al., 2005. High dose vitamin E therapy in amyotrophic lateral sclerosis as add-on therapy to riluzole: results of a placebo-controlled double-blind study. J. Neural Transm. 112, 649–660.

Haukedal, H., Freude, K., 2019. Implications of microglia in amyotrophic lateral sclerosis and frontotemporal dementia. J. Mol. Biol. 431, 1818.

Heneka, M.T., Kummer, M.P., Latz, E., 2014. Innate immune activation in neurodegenerative disease. Nat. Rev. Immunol. 14, 463–477.

Henkel, J.S., Engelhardt, J.I., Siklos, L., Simpson, E.P., Kim, S.H., Pan, T., et al., 2004. Presence of dendritic cells, MCP-1, and activated microglia/macrophages in amyotrophic lateral sclerosis spinal cord tissue. Ann. Neurol. 55, 221–235.

Igoudjil, A., Magrane, J., Fischer, L.R., Kim, H.J., Hervias, I., Dumont, M., et al., 2011. In vivo pathogenic role of mutant SOD1 localized in the mitochondrial intermembrane space. J. Neurosci. 31 (44), 15826–15837.

Ihara, Y., Nobukini, K., Takata, H., Hayabara, T., 2005. Oxidative stress and metal content in blood and cerebrospinal fluid of amyotrophic lateral sclerosis patients with and without a Cu, Zn-superoxide dismutase mutation. Neurol. Res. 27, 105–108.

Ingre, C., Roos, P.M., Piehl, F., Kamel, F., Fang, F., 2015. Risk factors for amyotrophic lateral sclerosis. Clin. Epidemiol. 7, 181–193.

Jagaraj, C.J., Parakh, S., Atkin, J.D., 2020. Emerging evidence highlighting the importance of redox dysregulation in the pathogenesis of amyotrophic lateral sclerosis (ALS). Front. Cell. Neurosci. 14, 581950.

Jaroonwitchawan, T., Chaicharoenaudomrung, N., Namkaew, J., Noisa, P., 2017. Curcumin attenuates paraquat-induced cell death in human neuroblastoma cells through modulating oxidative stress and autophagy. Neurosci. Lett. 636, 40–47.

Jiang, H., Tian, X., Guo, Y., Duan, W., Bu, H., Li, C., 2011. Activation of nuclear factor erythroid 2-related factor 2 cytoprotective signaling by curcumin protect primary spinal cord astrocytes against oxidative toxicity. Biol. Pharm. Bull. 34, 1194–1197.

Joe, B., Vijaykumar, M., Lokesh, B.R., 2004. Biological properties of curcumin-cellular and molecular mechanisms of action. Crit. Rev. Food Sci. Nutr. 44, 97–111.

Kaur, I.P., Bhandari, R., Bhandari, S., Kakkar, V.J., 2008. Potential of solid lipid nanoparticles in brain targeting. J. Control. Release 127, 97–109.

Khalil, N.M., do Nascimento, T.C., Casa, D.M., Dalmolin, L.F., de Mattos, A.C., Hoss, I., et al., 2013. Pharmacokinetics of curcumin-loaded PLGA and PLGA-PEG blend nanoparticles after oral administration in rats. Colloids Surf. B Biointerfaces 101, 353–360.

Kim, T.H., Jiang, H.H., Youn, Y.S., Park, C.W., Tak, K.K., Lee, S., et al., 2011. Preparation and characterization of water-soluble albumin-bound curcumin nanoparticles with improved antitumor activity. Int. J. Pharm. 403, 285–291.

Kim, D.S., Kim, J.Y., Han, Y., 2012. Curcuminoids in neurodegenerative diseases. Recent Pat. CNS Drug Discov. 7, 184–204.

Lacorte, E., Ferrigno, L., Leoncini, E., Corbo, M., Boccia, S., Vanacore, N., 2016. Physical activity, and physical activity related to sports, leisure and occupational activity as risk factors for ALS: a systematic review. Neurosci. Biobehav. Rev. 66, 61–79.

Liao, B., Zhao, W., Beers, D.R., Henkel, J.S., Appel, S.H., 2012. Transformation from a neuroprotective to a neurotoxic microglial phenotype in a mouse model of ALS. Exp. Neurol. 237, 147–152.

Liao, S.C., Hsu, W.H., Huang, Z.Y., Chuang, K.L., Lin, K.T., Tseng, C.L., et al., 2019. Bioactivity evaluation of a novel formulated curcumin. Nutrients 11, 2982.

Liu, W., Zhai, Y., Heng, X., Che, F.Y., Chen, W., Sun, D., et al., 2016. Oral bioavailability of curcumin: problems and advancements. J. Drug Target. 24, 694–702.

Lu, J., Duan, W., Guo, Y., Jiang, H., Li, Z., Huang, J., et al., 2012. Mitochondrial dysfunction in human TDP-43 transfected NSC34 cell lines and the protective effect of dimethoxy curcumin. Brain Res. Bull. 89, 185–190.

Lu, J.M., Lin, P.H., Yao, Q., Chen, C., 2010. Chemical and molecular mechanisms of antioxidants: experimental approaches and model systems. J. Cell. Mol. Med. 14, 840–860.

Lu, C.H., Allen, K., Oei, F., Leoni, E., Kuhle, J., Tree, T., et al., 2016. Systemic inflammatory response and neuromuscular involvement in amyotrophic lateral sclerosis. Neurol. Neuroimmunol. Neuroinflammation 3, e244.

Luss, A.N., Kulikov, P.P., Romme, S.B., Andersen, C.L., Pennisi, C.P., Docea, A.O., et al., 2018. Nanosized carriers based on amphiphilic poly-N-vinyl-2-pyrrolidone for intranuclear drug delivery. Nanomedicine 13, 703–715.

Mackenzie, I.R.A., Rademakers, R., 2008. The role of transactive response DNA-binding protein-43 in amyotrophic lateral sclerosis and frontotemporal dementia. Curr. Opin. Neurol. 21, 693–700.

Maiti, P., Dunbar, G.J., 2018. Use of curcumin, a natural polyphenol for targeting molecular pathways in treating age-related neurodegenerative diseases. Int. J. Mol. Sci. 19, 1637.

Marrache, S., Dhar, S., 2012. Engineering of blended nanoparticle platform for delivery of mitochondria-acting therapeutics. Proc. Natl. Acad. Sci. U. S. A. 109, 16288–16293.

McCombe, P.A., Henderson, R.D., 2010. Effects of gender in amyotrophic lateral sclerosis. Gend. Med. 7, 557–570.

Menon, V.P., Sudheer, A.R., 2007. Antioxidant and anti-inflammatory properties of curcumin. Adv. Exp. Med. Biol. 595, 105–125.

Mitsumoto, H., Santella, R.M., Liu, X., Bogdanov, M., Zipprich, J., Wu, H., et al., 2009. Oxidative stress biomarkers in sporadic ALS. Amyotroph. Lateral Scler. 9 (3), 177–183.

Moustapha, A., Pérétout, P., Rainey, N.E., Sureau, F., Geze, M., Petit, J.M., et al., 2015. Curcumin induces crosstalk between autophagy and apoptosis mediated by calcium release from the endoplasmic reticulum, lysosomal destabilization and mitochondrial events. Cell Death Discov. 1, 15017.

Mukhopadhyay, A., Basu, N., Ghatak, N., Gujral, P.J.A., 1982. Antiinflammatory and irritant activities of curcumin analogues in rats. Agents Actions 12, 508–515.

Naksuriya, O., Okonogi, S., Schiffelers, R.M., Hennink, W.E., 2014. Curcumin nanoformulations: a review of pharmaceutical properties and preclinical studies and clinical data related to cancer treatment. Biomaterials 35, 3365–3383.

Ojha, R.P., Rastogi, M., Devi, B.P., Agrawal, A., Dubey, G.P., 2012. Neuroprotective effect of curcuminoids against inflammation-mediated dopaminergic neurodegeneration in the MPTP model of Parkinson's disease. J. Neuroimmune Pharmacol. 7, 609–618.

Pan, J., Li, H., Ma, J.F., Tan, Y.Y., Xiao, Q., Ding, J.Q., et al., 2012. Curcumin inhibition of JNKs prevents dopaminergic neuronal loss in a mouse model of Parkinson's disease through suppressing mitochondria dysfunction. Transl. Neurodegener. 1, 16.

Pesaresi, M.G., Amori, I., Giorgi, C., Ferri, A., Fiorenzo, P., Gabanella, F., et al., 2011. Mitochondrial redox signalling by p66Shc mediates ALS-like disease through Rac1 inactivation. Hum. Mol. Genet. 20, 4196–4208.

Pickles, S., Destroismaisons, L., Peyrard, S.L., Cadot, S., Rouleau, G.A., Brown Jr., R.H., et al., 2013. Mitochondrial damage revealed by immunoselection for ALS-linked misfolded SOD1. Hum. Mol. Genet. 22, 3947–3959.

Rainey, N., Motte, L., Aggarwal, B.B., Petit, P., 2015. Curcumin hormesis mediates a cross-talk between autophagy and cell death. Cell Death Dis. 6, e2003.

Rakotoarisoa, M., Angelova, A., 2018. Amphiphilic nanocarrier systems for curcumin delivery in neurodegenerative disorders. Medicines (Basel, Switz.) 5, 126.

Roney, C., Kulkarni, P., Arora, V., Antich, P., Bonte, F., Wu, A., et al., 2005. Targeted nanoparticles for drug delivery through the blood–brain barrier for Alzheimer's disease. J. Control. Release 108, 193–214.

Shaikh, J., Ankola, D.D., Beniwal, V., Singh, D., Kumar, M.N., 2009. Nanoparticle encapsulation improves oral bioavailability of curcumin by at least 9-fold when compared to curcumin administered with piperine as absorption enhancer. Eur. J. Pharm. Sci. 37, 223–230.

Shanmugam, M., Rane, G., Kanchi, M., Arfuso, F., Chinnathambi, A., Zayed, M.E., et al., 2015. The multifaceted role of curcumin in cancer prevention and treatment. Molecules 20, 2728–2769.

Shaw, P.J., Ince, P.G., Falkous, G., Mantle, D., 1995. Oxidative damage to protein in sporadic motor neuron disease spinal cord. Ann. Neurol. 38, 691–695.

Shehzad, A., Rehman, G., Lee, Y.S., 2013. Curcumin in inflammatory diseases. Biofactors 39, 69–77.

Simpson, E.P., Henry, Y.K., Henkel, J.S., Smith, R.G., Appel, S.H., 2004. Increased lipid peroxidation in sera of ALS patients: a potential biomarker of disease burden. Neurology 62, 1758–1765.

Smith, R.G., Henry, Y.K., Mattson, M.P., Appel, S.H., 1998. Presence of 4-hydroxynonenal in cerebrospinal fluid of patients with sporadic amyotrophic lateral sclerosis. Ann. Neurol. 44 (4), 696–699.

Smith, E.F., Shaw, P.J., De Vos, K.J., 2019. The role of mitochondria in amyotrophic lateral sclerosis. Neurosci. Lett. 710, 132933.

Szymusiak, M., Hu, X., Plata, P.A.L., Ciupinski, P., Wang, Z.J., Liu, Y.J., 2016. Bioavailability of curcumin and curcumin glucuronide in the central nervous system of mice after oral delivery of nano-curcumin. Int. J. Pharm. 511, 415–423.

Tafuri, F., Ronchi, D., Magri, F., Comi, G.P., Corti, S., 2015. SOD1 misplacing and mitochondrial dysfunction in amyotrophic lateral sclerosis pathogenesis. Front. Cell. Neurosci. 9, 336.

Thiyagarajan, M., Sharma, S.S., 2004. Neuroprotective effect of curcumin in middle cerebral artery occlusion induced focal cerebral ischemia in rats. Life Sci. 74, 969–985.

Tiwari, S.K., Agarwal, S., Seth, B., Yadav, A., Nair, S., Bhatnagar, P., et al., 2014. Curcumin-loaded nanoparticles potently induce adult neurogenesis and reverse cognitive deficits in Alzheimer's disease model via canonical Wnt/β-catenin pathway. ACS Nano 8, 76–103.

Tohgi, H., Abe, T., Yamazaki, K., Murata, T., Ishizaki, E., Isobe, C., 1999. Remarkable increase in cerebrospinal fluid 3-nitrotyrosine in patients with sporadic amyotrophic lateral sclerosis. Ann. Neurol. 46, 129–131.

Tripanichkul, W., Jaroensuppaperch, E.O., 2012. Curcumin protects nigrostriatal dopaminergic neurons and reduces glial activation in 6-hydroxydopamine hemiparkinsonian mice model. Int. J. Neurosci. 122, 263–270.

Tripodo, G., Chlapanidas, T., Perteghella, S., Vigani, B., Mandracchia, D., Trapani, A., et al., 2015. Mesenchymal stromal cells loading curcumin-INVITE micelles: a drug delivery system for neurodegenerative diseases. Colloids Surf. B Biointerfaces 125, 300–308.

Tsai, Y.M., Chien, C.F., Lin, L.C., Tsai, T.H., 2011. Curcumin and its nano-formulation: the kinetics of tissue distribution and blood-brain barrier penetration. Int. J. Pharm. 416, 331–338.

Ullah, F., Liang, A., Rangel, A., Gyengesi, E., Niedermayer, G., Münch, G., 2017. High bioavailability curcumin: an anti-inflammatory and neurosupportive bioactive nutrient for neurodegenerative diseases characterized by chronic neuroinflammation. Arch. Toxicol. 91, 1623–1634.

Valdmanis, P.N., Rouleau, G.A., 2008. Genetics of familial amyotrophic lateral sclerosis. Neurology 70, 144–152.

Vargas, M.R., Johnson, D.A., Johnson, J.A., 2011. Decreased glutathione accelerates neurological deficit and mitochondrial pathology in familial ALS-linked hSOD1(G93A) mice model. Neurobiol. Dis. 43, 543–551.

Vogel, S.M., Minshall, R.D., Pilipović, M., Tiruppathi, C., Malik, A.B., 2001. Albumin uptake and transcytosis in endothelial cells in vivo induced by albumin-binding protein. Am. J. Physiol. Lung Cell. Mol. Physiol. 281, L1512–L1522.

Wang, X.S., Zhang, Z.R., Zhang, M.M., Sun, M., Wang, W., Xie, C., 2017. Neuroprotective properties of curcumin in toxin-base animal models of Parkinson's disease: a systematic experiment literatures review. BMC Complement. Altern. Med. 17, 412.

Weishaupt, J.H., Bartels, C., Pölking, E., Dietrich, J., Rohde, J., Poeggeler, B., et al., 2006. Reduced oxidative damage in ALS by high-dose enteral melatonin treatment. J. Pineal Res. 41, 313–323.

Wilkins, H.M., Morris, J.K., 2017. New therapeutics to modulate mitochondrial function in neurodegenerative disorders. Curr. Pharm. Des. 23, 731–752.

Xie, X., Tao, Q., Zou, Y., Zhang, F., Guo, M., Wang, Y., et al., 2011. PLGA nanoparticles improve the oral bioavailability of curcumin in rats: characterizations and mechanisms. J. Agric. Food Chem. 59, 9280–9289.

Yallapu, M., Jaggi, M., Chauhan, S.J.C., 2013. Curcumin nanomedicine: a road to cancer therapeutics. Curr. Pharm. Des. 19, 1994–2010.

Yallapu, M.M., Nagesh, P.K.B., Jaggi, M., Chauhan, S.C., 2015. Therapeutic applications of curcumin nanoformulations. AAPS J. 17, 1341–1356.

Yang, F., Lim, G.P., Begum, A.N., Obeda, J., Simmons, M.R., Ambegaokar, S.S., et al., 2005. Curcumin inhibits formation of amyloid β oligomers and fibrils, binds plaques, and reduces amyloid in vivo. J. Biol. Chem. 280, 5892–5901.

Yavarpour-Bali, H., Ghasemi-Kasman, M., Pirzadeh, M., 2019. Curcumin-loaded nanoparticles: a novel therapeutic strategy in treatment of central nervous system disorders. Int. J. Nanomedicine 14, 4449–4460.

Yu, B., Pamphlett, R., 2017. Environmental insults: critical triggers for amyotrophic lateral sclerosis. Transl. Neurodegener. 6, 15.

Zarei, S., Carr, K., Reiley, L., Diaz, K., Guerra, O., Fernandez, P., et al., 2015. A comprehensive review of amyotrophic lateral sclerosis. Surg. Neurol. Int. 6, 171.
Zhao, N., Francis, N.L., Calvelli, H.R., Moghe, P.V., 2020. Microglia-targeting nanotherapeutics for neurodegenerative diseases. APL Bioeng. 4, 030902.
Zorofchian Moghadamtousi, S., Abdul Kadir, H., Hassandarvish, P., Tajik, H., Abubakar, S., Zandi, K.J., 2014. A review on antibacterial, antiviral and antifungal activity of curcumin. Biomed. Res. Int. 2014, 1–12.

Index

Note: Page numbers followed by *f* indicate figures.

N

O

P

R

S

Made in the USA
Monee, IL
12 September 2026

62977355R00090